Total Quality

for
Safety
and
Health
Professionals

F. David Pierce

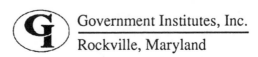

Government Institutes, Inc.
Rockville, Maryland

Government Institutes, Inc., 4 Research Place, Suite 200, Rockville, Maryland 20850

Library of Congress Cataloging-in-Publication Data

Pierce, F. David.

    Total quality for safety and health professionals / F. David Pierce.
          p.   cm.
    Includes bibliographical references and index.
    ISBN 0-86587-462-X
    1. Industrial safety--Management. 2. Total quality management.
  I. Title.
  T55. P54   1995
  658.4 ' 08--dc20   95-8268

                                  CIP

Printed in the United States of America

TOTAL QUALITY FOR SAFETY AND HEALTH PROFESSIONALS

CONTENTS

To my family. To Jeff, my son, who taught me how to enjoy life, my wife Gayle who teaches me how to live it, my son Kelly who's creativity gave this book its original title, and my other sons, Greg and Brad, who together with the other family members try to understand my passion for total quality.

FOREWORD: THE TOTAL QUALITY REVOLUTION

It's said that there are two certainties in life—death and taxes. Today, one more certainty can be added—change. It's true. In our world today, change is so unavoidable that in most instances we've become desensitized to it—desensitized, but not immune from its disruptive impact on our lives. No matter if the focus is on the traditional aspects of life such as children and home, or on the business world or technology, change is all around us, all of the time.

I grew up in the flatlands of New Mexico, not far from the Mexican border. There, the wind blows incessantly. We used to say that the airborne dust in our living room naturally went from south to north across the room. Natives joke that the wind is so constant that if it ever stopped, people would immediately respond, "What's that?" as if a sudden noise had shattered the quiet.

It's the same for change today. If change were stopped, we would question with alarm, "What's that?" as things would definitely not be normal.

Change is both good and bad. Good because it adds excitement and learning experiences. A friend of mine can hardly wait each time a new advancement in electronic gadgetry is announced. He is excited by change. I, however, still haven't learned how to program our two-year-old videocassette recorder. First, it seems too complex. And second, why bother? Next year a new one will take its place and I would have to unlearn the present recorder's workings and learn another. I have to admit, though, all those new features and additional lights are pretty exciting.

Change also has a dark side. It can be disruptive. Humans are patterned animals—we like predictability. We depend a great deal on predictability in our lives for our mental and physical well-being. Nature, for the most part, is predictable. Winters are colder, springs are wetter, summers hotter, and autumns are the prettiest. Going home, the greetings, the surroundings, the people encountered generally are predictable. When this predictability is disturbed by change, invited or not, disruption results. The unavoidable result of disruption is stress.

The amount of stress is directly related to the magnitude of the disruption and how long we will be affected by the change. The greater and longer the disruption, the larger the applied stress. For example, you can live with the disruption of your mother-in-law visiting if you know that she will be leaving soon. If she moved in, however, the stress level would significantly increase. Understanding the magnitude of disruption is a little more complex. Here, the study of alcoholism provides answers. Psychologists know that an alcoholic, as a result of his or her illness, will sacrifice spouse, family, friends and possessions before he or she will risk losing their job. So, the amount of disruption is potentially greatest at work.

Think about it. If something changes significantly at work, it can disrupt your entire life. It is important to realize that never before have America's workplaces undergone such rapid change: workforces are being downsized; management structures are being drastically altered and streamlined; facilities are being closed, enlarged, or modified all the time; new job responsibilities and challenges are common; functions are being eliminated or out-placed; new and very different ways of managing are being tried; traditional success formulas are being discarded.

One of the most significant changes at work has been termed the "Quality Revolution." Definitions and expectations of quality have been drastically altered and processes to bring about greater efficiency and quality are springing up. Out of this Quality Revolution has emerged a significant management concept that has been termed Total Quality or Total Quality Management (TQM). Compared to our traditional ways of managing in America, total quality is different, very different. Therefore, a management change to total quality is disruptive, very disruptive.

The safety and health profession is not immune to these changes. As any change occurs, though, some parts move faster than the rest. Communication between those who have made the change and those who haven't becomes difficult. It results mostly from a difference in perspective. It's one of the accompanying problems that comes with starting to think differently. Because of different perspectives, or mind sets, people simply have difficulty communicating about change. The important point is that communication difficulty actually impedes the change process. This communication problem is a lot like being a member of a remote central South American Indian tribe trying to get directions from someone in New York City.

This isn't a newly discovered phenomenon. For example, look at the Vietnam War protesters of the late 1960s. Opinions were rapidly changing. The perspective of both the protesters and those in America's government came from two different directions. What happened? They couldn't communicate. As a result, the protesters were convinced that the government and military were composed of nothing more than war-mongering imperialists intent upon raping the land and killing innocent civilians. On the other hand, the government's opinion of the protesters was expressed in words such as "nonpatriotic" and "cowardly." Neither side could comprehend the other's point of view because of their different perspectives. Because neither could communicate with the other, no resolution could be found and no common ground discovered.

American industry is another example. Consider the long-running communication problems that exist between management and labor—especially evident in unionized operations during contract negotiations. Management's position typically is that labor wants too much and management can't afford it. Labor's perspective is very different. Labor feels that they deserve more for what they do. If the business were managed more effectively, management could afford it. Reality? Who is right and wrong? It doesn't matter. Their different perspectives, however, do. They make communication next to impossible.

Another example of communication problems that happen because of different mind sets or perspectives is close to a lot of us—a parent and a teenager. Do parents have difficulty communicating with their teenager? Does the teenager have difficulty talking to his or her mom and dad? Of course they have difficulty communicating—not because they speak different languages, but because they have different perspectives.

So, this book is about change and we, like a parent and his or her teenage son or daughter, start with a strong chance of having some communication difficulties because of our different perspectives. Please don't lose patience and quit trying. Within this book, we will walk together along a fascinating but perplexing path. It is rife with snares, holes, low branches, and strange creatures. Most of them may be very difficult to see, question, or accept. We are beginning with a different mind set. I no longer accept many traditional concepts that I was taught. The times are changing. And we as safety and health professionals, like American industry, must challenge traditional methods and try new ways of thinking if we are to be successful in the future. The solutions of yesterday may not serve us in the future. Those who believe that safety and health concepts and concerns of yesterday are the same today—and will be so tomorrow—may not succeed. Until these traditional safety and health concepts are challenged, our profession and purpose will struggle, fail, or be assigned to those who can think differently. Changing isn't a win–lose concept. It can be a win–win future!

This challenge, therefore, is presented to us as safety and health practitioners: widen our perspectives so that we do not immediately reject new ideas. In other words, we need to look at new ideas like total quality with an open mind and withhold judgment. My challenge is to communicate to you in "friendly" language using many examples so as not to induce "brain rejection" too soon in the questioning process.

You might think that the purpose of this book is to change the perspective or culture of the safety and health professions. That ultimately might be correct; in reality, though, the purpose is much simpler. In fact, the purpose is to cultivate a personal change in you. I recognize that changes of any kind are neither easy nor quick. I also recognize that the only way to change the profession is by making one "convert" at a time. As you question safety and health and total quality, you can be a champion, supporter, or leader of change. And change is not only focused in the workplace, but also in the wider context of the safety and health profession.

As you read this book, I ask five things of you. First, as you face concepts or positions that run counter to your perspective, identify your resistance as quickly as possible. Understand that resistance is a perspective-induced reaction only. Second, don't become protective of your perspective or defensive until you fully understand my perspective. Immediately becoming defensive will hold back or stop your learning process. Third, as necessary, accept some ideas on "faith" until you can see all sides. Then, accept or reject them as you wish. Fourth, actively participate in the learning experience. Take notes about those ideas you agree with, disagree with, have additional questions about, or need to think about more. And fifth, regularly question your purpose in reading this book and your perspectives as you read. It is through your regular questioning that you can learn more about yourself and augment what I have to offer.

Three different results are possible from reading this book. First, you could reject everything. That's your prerogative. You could decide that total quality has no validity or application. Second, you could learn how to put total quality into your safety program. Many important concepts and tools described in this book can notably improve any safety and health program. And third, you could learn how to put safety into total quality, which is my ultimate purpose in writing this book.

If the first outcome should result, we both lose. I lose because I did not communicate in a manner that met your needs. You lose because you have not questioned or learned anything. Every experience in life offers an opportunity for learning. If the second result occurs, then I partly win by getting you to question some of your thoughts about safety and provide you with some useful information that can improve your safety and health program. You also win in that these tools can decrease your frustration and burn-out, and ultimately elevate your safety program to a higher level. If, however, the third result happens, we both win. The real winner, however, is the safety and health profession. An important event will have happened. We will have begun the process of changing the culture of our profession.

We are going to talk a lot about the concept of total quality or TQM. It goes by many different names and acronyms. Notice that I called it a concept. Total quality is not like describing an aardvark. It looks very different in different applications and from different perspectives. *Successful* total quality programs do, however, have common concepts and components, and those will be discussed in great detail. The parts you use are up to you and your particular application. Information about the continuing evolution of total quality and total quality tools is also included. Information on roadblocks is offered and how the process of change happens. So, between the covers of this book, I've tried to give you as complete a reference as possible.

The world is in the grips of another revolution, not only in industry but in business as we know it. The workplace is changing at rates never thought possible. American industry is changing out of need. Business may strive to become more lean, more profitable, more efficient. In some areas, competition is driving this change. In others, public demands or funding reductions are fueling changes. No matter what triggers the change, the important realities are that change is occurring, it is happening rapidly, and business as we know it will continue to change.

In industry, such change has been called the "Quality Revolution," but it is much deeper and broader than that. Many descriptive terms are being used, one of the most popular of which is Total Quality Management. TQM by itself, refers only to this change. Too often we use it inappropriately to describe a certain way of doing business or to describe a business culture or philosophy. "We're a TQM company." The term just isn't proper for these uses. Total quality has not only changed the way business thinks but also how it operates; it will continue to do so because total quality is a concept that describes the changing business environment, not a specific process by which change occurs.

Years ago, as American industry was beginning its voyage into the total quality waters, I began to tell both my professional and management peers and those in service-type businesses about the possibilities it offered. Surely the advantages of total quality were obvious and they would choose to change with industry, thus pioneering the change in their field or business. I felt that the total business environment after the year 2000 would be notably different from what we had seen in the past. Over the past few years, I have seen that I was right. Today, there are two prominent camps. Those who have changed their perspective or culture to total quality, and those who have not. I include many businesses that only give lip service to total quality in the group of those who have not changed their perspective or culture. It's in vogue to say that you are a total quality business but, in reality, to have made only small or superficial changes.

Let me tell you of a recent experience that helps illustrate this point. I was preparing to teach a total quality course. A corporate manager of safety from a Fortune 100 company called me about attending the course. She was recently appointed to the position and was seeking tools that she could use to build an effective program. Total quality was an obvious need to her. "I talked to my boss about sending me to your course," she said. "My boss doesn't feel that it will help me. He says that I can learn all I need about total quality here." "What do you think?" I asked. "Well," she began, "from what I've seen, if they know what total quality is, I haven't seen it yet." Unfortunately, this is a very common occurrence. To date, I've talked with about twenty top safety and health professionals in Fortune companies who share this confusion and frustration.

The concept of total quality requires a total change in perspective or thinking. It is like converting to a certain religion. If you profess that you have converted and show you have by performing some of the actions, but do not change "in your heart," conversion has not occurred.

Changing to a total quality perspective has some verifiable evidences of "conversion." An easy way exists to identify those who have and those who haven't. They lose the ability to communicate with each other. Those that have not changed perspectives or culture continue to speak in classical "business-ese," but inject total quality words. Those businesses that have "converted" to total quality take on a new language. I call it "future-ese." Conversing from either side is equally frustrating. Although they use some of the same words, the language is different. Seen from the classical business perspective you would hear, "I see your lips move. I even think that I understand your words but it gets all garbled-up in my mind. I don't understand at all." It is not unlike reaching "tilt" on a pinball machine. The process stops, the communication pathway shuts down. From the total quality perspective, it's equally frustrating.

I have three teenage boys at home. At times they are a real source of love, excitement, and pride. At other times, they are a source of great frustration. Recently, I was trying to help one of them with his algebra homework. From my perspective, the problems he was struggling with were easy to do. From his, they were anything but easy. I had confidence that I could show him how to learn the process by showing him and talking him through it. After twenty frustrating minutes, I lost patience and gave up. Ahead of me, he became frustrated and quit listening and trying. (He's quicker than I am at recognizing an impasse.) My oldest teenage son came to my rescue and was successful at helping his younger brother.

Why did this happen? My son and I were both coming at the same problems from different perspectives. And because of those different perspectives, we had difficulty communicating. On the other hand, both brothers shared perspectives *and* language. As in this example, total quality converts and "nonbelievers" have the same difficulties communicating and the same associated frustrations.

So, where do we begin our journey together? We start at the beginning. We need to discover what this concept called total quality is all about. All you need on this journey is an open, questioning mind to discover the rewards and satisfaction of total quality. Along the path we will meet frustrations, communication challenges, and new ideas; if we complete the journey, we will succeed together and begin this important change.

*T*otal *Q*uality

for Safety and
Health
Professionals

1

WHAT CAN TOTAL QUALITY DO FOR ME?

"We're a TQM company," said a business colleague to me. He continued, "In today's world, there just isn't any other way to operate. Besides, TQM allows us to be so much more productive." If he had just stopped with the first declaration, I probably wouldn't have even thought of probing. Now it was irresistible. "Oh, so your company's into total quality; what do you measure?" "We measure everything," he replied. "We measure production, on-time shipments, errors, and other business indicators like that. As a matter of fact, our walls are full of graphs. It's really impressive!"

"What do all your graphs tell you?" I asked. "Well, they tell us that we're getting better, a whole lot better," he answered. "Better at what?"

"Do you have any idea of how much our production has increased over the past three years?" he responded. "No, I really don't have a clue." "Well, we've increased our production by more than 22 percent."

"How is your error rate doing?" I asked. "Error rate? Oh, you mean product rework? Well, that's about the same. We can see some real strong indicators that we are gaining on it though."

"How's your work force?" I continued my probing. "Work force? Well, it hasn't grown as much as our production has. That alone has to say something about the efficiency of our production, doesn't it?"

"What's your corporate vision and what are you benchmarking against?"

"First, we want to be the best. We will accept no other position. Benchmarking? I'm not sure I know what you're talking about. To my way of thinking benchmarking is another way of saying measurement."

Now that was an interesting answer! "Has your safety staff increased? I questioned. "Oh sure, but with the new regulations, we need more people to do the work," he replied. "Can't the line structure absorb the responsibilities?" I asked. "Are you kidding, they can't do our job. Besides, we have to constantly oversee what they do or it never gets done," he snapped back his reply.

"Sounds like kingdom building to me," I responded quietly. He did not catch what I had quietly muttered.

It's a pity. I was just warming up. "I was just curious," I said and ended our conversation. It was obvious at that point that total quality to him and to his company was a reincarnation of MBO, Management by Objectives. It wasn't total quality at all. It was only in their minds and mostly made of "smoke and mirrors." But, they did use those total quality buzzwords!

TOTAL QUALITY: USEFUL CONCEPT OR FAD

It's a real shame because this is all too frequently what is found when searching for total quality companies. In business today, there is great confusion about what total quality is and what it can do. If we are to discuss total quality in this book, knowing what it is is pretty important. As one person I talked to about total quality so eloquently said, "What the hell is total quality anyway?" Good question and a good starting point.

All fads and rages in our society manifest themselves in many different ways. If it's "In," everyone simply has to have it, whether the fad is clothes, nightlife, imported cars, or wristwatches. Total quality is not an object. It's a process. And, that's where the "everyone has to have one" concept derails. It does so because, in fact, everyone *doesn't* have to have total quality. A company can choose to conduct "business as usual" instead. In my opinion and in the opinion of an ever-growing group of people, not embracing total quality will mean the difference between in-business today and out-of-business tomorrow. It is that important! In reality, total quality today is perceived in business simply as a fad. Any person or any company of consequence just has to have it. It's kind of like having marble building foyers or glass elevators. It's "In," so everyone has to have it.

TOTAL QUALITY: WHAT IT ISN'T

"Tell me. What is a total quality program?" I asked a manager in a Quality Department. He responded, "Quality programs define standards and specifications for products, are documentable, and auditable. Total quality programs emphasize, in a verifiable way, that the approach to quality is documented, its implementation is trackable, and the results are a product of that approach and successful implementation." Wow! Is *that* what total quality is? *Not!* This "quality person," as do so many, tried to stick total quality into a bottle so it could be defined. It's an analytical mind's dream! This firm vision of quality is continually reinforced by the presence of Quality Control Departments and by the quality competitions including the national quality competition, the Malcolm Baldrige Award. Total quality *cannot* be stuck inside a bottle for inspecting or auditing. Total quality is *much* bigger than that. It's no wonder there are so many misinterpretations of what total quality is and what it can do!

Let's get away from trying to put total quality inside a bottle. Let's look at it from a larger conceptual perspective. With all the misinterpretations and misuses of total quality in today's world, we might best begin by discussing what total quality is *not*. Keep in mind, total quality is *none* of these. First, total quality is *not* MBO revised. Management by Objectives was a management concept that tried to bridge the gap between management by luck, and management by plan. Originally conceived in the 1950's, MBO became a fad, much like total quality is considered to be today. MBO is nothing more than planning what you want to accomplish and measuring your success (or failure). Although, successful total quality programs do utilize extensive planning and performance measurement, total quality takes them a big step farther.

Total quality is also not new generation Theory Y management. Theory Y management has been called management by "huggers and kissers." In this theory of management, subservient managers allow workers to find problems and fix them. Management, therefore, serves the employees and allows them to accomplish improvements in the workplace. It is true that successful total quality programs have high employee participation, but management is far from subservient. In a total quality program, management must lead the efforts. Therefore, total quality is definitely *not* new generation Theory Y management.

Conceptually, total quality is commonly thought to be merely management according to a strategic plan. This is an expansion of the MBO concept of planning, but it is longer ranged and multifaceted. In successful total quality programs and companies, strategic planning is an integral part of the process, but only a part. Describing total quality as strategic planning, however, reduces the process to only one of many tools used in total quality programs. Therefore, total quality is also *not* management by a strategic plan.

Total quality is sometimes considered to be a manufacturing concept only—with no applicability to either service industries or staff functions. No, total quality is bigger than that. Remember: total quality is not an object. It is a process, and therefore can be applied to any business or industry. It can also be applied to any function including staff functions, production, or even running a Boy Scout troop. True, total quality has been mostly applied in the manufacturing arena because of manufacturing's need and willingness to change. It is not a function of applicability. Total quality is, therefore, *not* just a manufacturing concept.

Lately, total quality has been equated with statistical process control. With all the measurement used in total quality programs, some think that statistical process control must be the key. No, there are two problems in the equation. First, measurement in total quality has to be easy and easy to understand. If you work too hard on measurement, you waste valuable time. Statistical process control is complex. Second, measurement is only used in total quality to document improvement or accomplishment. It is a total quality tool, not the whole banana. Total quality is *not* statistical process control.

Finally, a newer misinterpretation goes something like this: TQM=ISO 9000. ISO 9000 is an object, a *long* list of quality specifications. Total quality, on the other hand, is a process, a way of thinking and doing business. Total quality is *not* ISO 9000.

Total Quality is *NOT*:

- **MBO Revised**
- **New Generation Theory Y**
- **Management by Strategic Planning**
- **A Manufacturing Concept ONLY**
- **Statistical Process Control**
- **ISO 9000**

So what *is* total quality? Society is driven by definitions. If it can't be defined, it must not exist. Here is a simple definition of total quality: a systematic, highly participative process that identifies inefficiencies and improves them. In reality, total quality is no more complex than that. Adding more specifics to total quality only limits its application or focus. Total quality is a process that is fueled by a different way of thinking. In a holistic sense, it can be defined even more concisely. Total quality is a highly participative process for improvement. This is why there is so much confusion over total quality. It's *too* simple! Because of its power, it is natural to make it more complex, but in reality, it is not.

Total Quality Is a Highly Participative Process for Improvement!

So, if total quality is so simple, what makes it so important and powerful? It is important because it is the way of the future. It is powerful because it changes cultures, personal and business, and the process by which improvement is achieved continually renews itself by the challenges, participation, and successes. Look at all the examples of companies that have implemented total quality programs. Look at their successes—higher production, lower costs, reduced inventories, higher quality, high employee morale, higher profits. It is difficult to argue with examples like Hewlett Packard and Motorola. Would they be as successful today if they had not risked their futures on total quality? Maybe. But most likely not. Sure, the market, good employees, and dumb luck might have produced some positive results, but not of the magnitude or the sustained performance that total quality has produced. Powerful? Yes. To most of America's companies and endeavors, however, especially in the staff and service sectors, total quality is still an untapped resource.

TOTAL QUALITY: WHERE DID IT ORIGINATE?

Like most concepts and processes, total quality doesn't have a single originator. Rather, it has many roots. The most important one, however, was an American, W. Edwards Deming. After World War II, Deming openly offered his new manufacturing concepts to American industry. It wasn't interested! And why should it have been? American companies represented the strength of the world's manufacturing. They were, in fact, the champions. They had no reason to change what had gotten them to that point. "If it ain't broke, don't fix it!" Did Deming give up? In frustration, he went to Japan and taught them Japanese. After the war, Japan had nothing to lose and everything to gain. And with Deming, gain they did.

The Japanese learned so well that I have separated their contribution to total quality from that of Deming. Anyone who has studied the Japanese or Japanese industry knows all too well that one of the strengths of the Japanese is their "learning culture." They can study something and learn it so well that they improve it to near perfection. That is just what they did with Deming's manufacturing ideas: expanded, refined and perfected them. The Japanese have contributed significantly to the development of total quality.

European industry also helped mold what we call total quality today. Europeans and their industries have long had a "quality is everything" attitude. Brands such as Mercedes, Porsche, Hasselblad, Rolex, and BMW are standards of quality. In fact, all manufacturers, including the Japanese, compare their product quality to them. Unfortunately, Europeans have mistakenly held to the idea that a knowledgeable consumer will always pay a high inflated price for high quality. In fact, with what the Japanese have brought, the idea that quality can be attained at reasonable costs, the European approach is being aggressively challenged today. Nevertheless, European industries made important contributions to total quality as we know it.

Total quality was not totally born abroad. Some of its development and processes are home-grown. For example, during the 1970s, self-managed teams swept America. When this concept fell short of the mark in making the changes that were anticipated, most industries and facilities quickly discarded it. However, one of the side effects of self-managed teams took root and is now an integral part of the total quality process—employee participation, input, and ownership. What started out in American industry as a good idea with limited application provided a building block for total quality today.

Total quality has also been significantly affected by the world's technological revolution. Yesterday, product research, development, design, and, finally, manufacture took many years. In our fast-paced, highly competitive world, this same process today takes only months. In some cases, it takes only weeks to go from concept to consumer. The technological revolution has had enormous impact on the development of total quality. This speed demands that communication and accuracy become woven into the total quality fabric. Just like today's concept-to-manufacturing cycle, total quality has had to become very fast without losing quality or value.

In fact, total quality today is a blend of many cultures, thoughts, industries, people, and revolutions, and it will continue to evolve. It is a dynamic concept, not static. Total quality tomorrow will certainly be much different from what it is today.

What do we know about total quality at this point? Basically, other than a simple definition, we know five things about total quality. First, conceptually, total quality is a marshmallow. Whereas successful total quality programs have several concepts and aspects in common, an all-inclusive definition of total quality is impossible. Second, total quality is different in most applications. Understanding that it is not an object but a process goes a long way to make this point clear. Third, total quality is a "moving target." By definition, it is always improving and, thus, always changing. Total quality tomorrow will be better and stronger than it is today. Fourth, total quality is *the* business concept for the twenty-first century. Businesses that are successful and survive will either consciously embrace total quality or stumble into its tenets by dumb luck. And fifth, those who succeed in total quality will be the visionary leaders of the future. Blind followers or those who ignore it will become history.

THE DISCONNECT

Several years ago, I had a particularly frustrating experience. The security contractor that serviced our facility functioned under classical Theory X management. Having its roots deep in the military, the management style of security businesses has strayed little from that authoritative, dictatorial style. Our company had changed our way of thinking to total quality and, as expected, the security contractor's management and I began to have trouble communicating. Excited about what we were experiencing at my company, I wanted to share it with them. But, because of our different perspectives, that was very difficult. One of the things that fueled my excitement was the win-win opportunity presented. If I could somehow change the way they managed their service and the people assigned to our facility, and instill some total quality concepts into the way they did business, I would receive not only vast returns in the quality of security we received, but also experience less turnover caused by their military-type management. The other win would be theirs. Having total quality, they would pioneer the change to it within their own industry and place themselves in a leadership position for the future.

What happened? Well, after about two years of efforts, I realized that they would not be able to make the transition to total quality, and we changed security contractors to one that was more receptive. Looking back, I recognize that we had two huge roadblocks to our success—language and perspective. For example, when I would speak of challenging people to get involved and take personal ownership in improving their security function, their interpretation of what I said translated like this. "Men (some were women), you need to work harder and stop screwing up." No wonder the assigned security people were confused and depressed. I wanted to improve the function, but from their management, they knew only that they were failing.

The second disconnect was our difference in perspective. I knew that total quality was a "plum" that was waiting to be picked and could be immediately planted into their industry with great success. They, however, looked at total quality as a manufacturing concept and could not see how a square peg could be driven into a round hole. To them, total quality was not applicable to their business. They would have had an easier time dealing with my insisting that they cross-dress. Their perspective of what total quality was applicable to kept them at arm's length from its advantages. We all ended up frustrated.

Their perspective that total quality is only a manufacturing concept, however, is a common misconception. Why? Well, mostly because total quality has been pioneered in the manufacturing sector. Thus, everything you read describes total quality from that perspective. Because we "buy into" that perspective about total quality, it is easy to discount it as a viable process for other applications, such as in service industries or safety and health programs.

It's a lot like our perspective of adhesives. What are adhesives used for? They stick objects together so that they don't come apart, right? Well, a few years ago our perspective about adhesives changed. It happened because a researcher at 3M discovered an adhesive that wasn't very good at sticking things together. Was it a failure? No, because 3M was able to change our perspective of adhesives, the Post-It Note® became an overnight success. Perspectives are extremely important and can limit us only if we let them!

TOTAL QUALITY IN A MANUFACTURING SETTING

In taking a preliminary look at what constitutes successful total quality programs, let's look at manufacturing total quality and see what it offers. Manufacturing total quality has some basic concepts that are integral in successful programs. The first of these concepts is focus. Successful total quality programs are both internally and externally focused. They focus internally at continually improving or getting better. They focus externally at those other programs, companies, or facilities that are leaders and from which they can learn.

The second concept in total quality is high-level participation and communication. Employees are involved in the improvement process at all levels of the organization or facility. That participation is fueled and reinforced by active multidirectional, multipathwayed communication.

Third, successful total quality programs are based on constant learning. They learn from history and from successes and failures, both internal and external. Their learning is focused on continually building skills and knowledge within the organization.

The fourth concept is the knowledge of, and commitment to, making small changes rather than large ones. Unlike classical American manufacturing thought, total quality companies know that small changes net more than large ones.

Fifth, successful total quality programs operate by dynamic planning and measurement. Planning and measurement are not merely activities to them, they are their pulse.

The sixth concept is risk taking and knowing that tomorrow will be discovered at the edge of today. Off-the-wall ideas can be *the* method of the future.

And finally, successful total quality programs know that improvement *never* stops. It's a continuum. If improvement stops, so will the effort or business, as it will be passed by those who continue to improve.

This last concept is extremely important within total quality. Worded in other ways, "You never arrive," "Your journey is never over!" Total quality is a moving target, always just disappearing over the horizon, always being pursued, always just a step away. In total quality you never "make it," because the standards and expectations are also constantly changing. Previously unthought of efficiency, production, and quality becomes possible and even obtained in a constantly improving environment. For example, years ago, simple computers filled rooms and took considerable time to do detailed, but simple tasks. Through continuous improvement, today's computers sit in the palm of your hand and work at blinding speeds. Twenty years ago, even ten years ago, that would have been considered impossible!

Think of the constantly changing face of successful total quality programs from the classic "Jack and the Beanstalk" perspective. In a total quality beanstalk, the climb is endless as the beanstalk continues to grow as we climb. This is one of the major reasons total quality is not for the faint-hearted or those who are involved in the "program of the week" styles. It is also a major reason many companies discard total quality as unworkable. They simply lose patience and focus on the long pathway. Total quality isn't something you "stick your toe into" to see if you like it. You have to jump in, whole body, all or none, and *then* stick with it.

TOTAL QUALITY IS CHANGE

Once there was a young boy who bragged that his dad was the strongest person in the world. Whether it was true or not, the boy felt it to be so. The boy had a favorite tree in his backyard. He would lay on a particular branch, high in the tree, and daydream. It was his "secret place," a place where he was extremely comfortable and felt safe. From his perch, he could see everything that was going on. He used to pretend that he was on "look out" for would-be robbers and thieves. One day, as he lay on his branch daydreaming, he became so inwardly focused that he did not see the weather changing. The sky grew darker and thick clouds appeared. The gusting winds quickly brought his thoughts back by almost lifting him off his branch. With his shirt blown up to his armpits, he took firm hold of the branch. He wrapped his arms and legs tightly around what used to be his quiet sanctuary. The branch began to pitch violently up and down, side to side. He began to panic.

"Help, Daddy!" The boy yelled. "Help!" Between the violently tossing branches he saw his father walk to the back window and look out. "Help, Daddy!" He yelled again. "Daddy, up here, hurry!" *CRACK!* The branch the boy was clinging to began to break. *CRACK!* The boy could feel the branch beginning to give way. "Help, Daddy! The branch is breaking! Help!" *CRACK!*

Below the boy, his father appeared. The boy knew that his father would climb the tree and rescue him. Much to the boy's shock, though, his father stayed on the ground beneath him. "Jump!" His father yelled up to the boy. What! And let go of the branch? *NO* way! Never had his father's arms looked so skinny and weak. "He could never catch me," the boy thought. "Jump!" His father again pleaded. *CRACK!* It was too late. The base of the branch finally splintered and down it came, boy and all, plummeting toward the earth. "Gotcha!" his father said triumphantly as he caught boy and branch. The boy wrapped his arms and legs tightly around his dad. Once again, his dad was the strongest person alive!

What does this simple story teach us in our quest for total quality? American business, as well as you and I are a lot like that boy. We lie comfortably in our lofty "look out," confident that we can see all foes coming our way. Complacent, we doze off and don't see the storm of changing business concepts brewing. We might see it beginning, but after becoming complacent, we continue to daydream. Soon that storm will begin to shake our comfortable perch. In some ways, the violent winds have already begun. Pitched by those winds of change, now we face a dilemma. We can ride out the storm clinging tightly to our old ways in the hope that they will not be destroyed by the storm and send us plummeting to the ground. Or we can call for help. The trick, however, is knowing who or what can catch us. Total quality can be that strong-armed savior that can cushion our fall. We need to have confidence that it can catch us. Better yet, with the help of total quality, we can build a strong fortress in our tree that will weather the storm. Most of us *are* a lot like that boy, but to some of us, the branch has already begun to crack! Hear it?

In summary, total quality is *the* concept, *the* process of the future. Don't try to define total quality and place it into a handy, understandable box. By doing so, total quality's scope and what it can achieve is limited. Don't mistakenly assume that it is merely a resurgence of some management concept from the past. Total quality may use many successful past concepts within its tapestry, however, it is not definable by any of them alone. Don't be duped into thinking that once you become a "TQM Company" that you've "made it." Total quality is a process for improvement. In fact, you *never* "make it." And finally, don't think that you can wait any longer to see if total quality will merely die as a fad. The world is changing rapidly. To paraphrase a popular saying, "In today's world, you can either lead (by embracing total quality), you can follow (and fall further and further behind), or you can get out of the way (step aside and lose your chance at success in the future)." Total quality isn't an option for success. It's *the* formula for success!

2

WHY ARE PARADIGMS IMPORTANT TO TOTAL QUALITY?

Paradigm. It's one of those "In" words, one now in vogue. We hear it on television and on radio. It's in newspapers and magazines. Some articles even become passionate about it. Usually, the point or question is whether or not a "paradigm shift" has occurred. Most authors appear to be in love with the word. But the importance of the concept is lost.

What is a paradigm? Where on earth did it come from? Why is it important to us as we discuss total quality? Good questions to explore.

The first time that the word paradigm was used in print was in a 1962 book by Thomas Kuhn, a scientist at the University of Chicago. In his book, *The Structure of Scientific Revolutions*, he discussed the dynamics of dramatic change and advancement in science. In his book he introduced the term paradigm as a potential roadblock to new ideas. Paradigms are beliefs that run so deep that they can blind a scientist from seeing new revealing data. As you would probably guess by Kuhn's subject matter, only a few people might pick up his book and struggle through the intricate language and detailed scientific reasoning. Consequently, the term paradigm remained buried deep inside its covers for many years.

Along came Joel Barker, a futurist. Joel Barker had not only read Kuhn's book and understood it, he clarified the term "paradigm." In his training packages he had a much friendlier vehicle to help popularize the term, the videotape. It also appealed to a much wider audience, business. In his videotape program, *Discovering the Future: The Business of*

Paradigms, Joel Barker caused a one-man explosion of the term in 1989. Barker concluded that a "paradigm" was a set of internal rules that did two things: 1) established boundaries, and 2) taught how to be successful by solving problems within these boundaries. Thus, paradigms work as information filters. Information that agrees with the paradigms is easily accepted, but information that does not is usually rejected and, in some cases, not seen at all.

Since that time, the word paradigm has taken on a life of its own. Unfortunately, with this popularity, it is also used to mean a bunch of different things. It's been used to mean a direction, a misconception, and even a measure of "bull-headedness." Those who have paradigms are cast as idiots or enemies of progress. Those who actively and passionately attack paradigms are conveyed as rescuers of the "American Way." In reality, if one uses the term as it was intended, it is neither negative nor positive. It's like having a spleen, it merely exists. And factually, paradigms aren't something that only other people have. They exist in everyone, including you and me.

Look at paradigms this way. Your broker tells you that you should consider investing in XYZ Corporation. He tells you that it will be a strong investment and will do well for you. You, however, have been doing some reading about where to invest your money. From what you understand, XYZ Corporation is a terrible investment! Therefore, you choose not to take your broker's advice. What both of you have are paradigms concerning XYZ Corporation. You have a strong belief that it is not a good investment; your broker, as a result of his sources, has a paradigm that is 180 degrees different from yours. Are you or your broker bad because each of you have a paradigm? No. The important lesson here: don't make the concept of the paradigm too big. It is a simple concept. It's what paradigms can do or keep us from doing that's important, especially when they get in the way of making good decisions or seeing important changes and new concepts.

PARADIGMS: CAUSE AND EFFECT

In a recent accident investigation at our company, we chased the causes of an injury through a maze of different conditions and actions. The investigation committee finally concluded that a series of unsafe acts had resulted in the injury, a chain reaction. No one could argue with the findings. However, the important disconnect came when the committee offered its recommendations to prevent recurrence. The recommendations were based on one assumption—that everyone knew that the actions were unsafe. In reality, there was a paradigm problem that went undetected by the investigating committee. This paradigm problem greatly limited the recommendations' success. You see, the people who were involved in the series of unsafe acts, and, subsequently, the injury, had a paradigm that was different from the investigating committee's. Recognizing "hindsight is always 20/20," the committee looked at the actions as unsafe and, thereby, unacceptable. Why? Because, someone got injured! Take away the injury and the outside investigating committee and you are left with the paradigm that was held by the work group. What was the workers' paradigm? They felt there was *no* safe way to do this type of job. So, as part of their jobs, they routinely accepted some unsafe actions as their best possible options. In the workers' paradigm, they didn't do the job unsafely. *They did it as safely as possible.* Therefore, the actions that caused the injury were *not* unacceptable. They were merely necessary to do the job. This was a wide divergence cause by different paradigms.

Why are paradigms important in our total quality search? Look at a paradigm as a mental barrier that defines reality. Paradigms, therefore, have three effects. First, they mask recognition. As Barker demonstrated in his video, it's like seeing a red-spade playing card in rapid fire. Your mind tells you that spades are black. If asked, a majority would not have perceived what they saw because it ran counter to their Playing Card Paradigm. Second, paradigms limit learning. This is a step beyond recognition. Paradigms define reality. Too often, when something comes and disagrees with our concept of reality, it is simply discarded it as a data outlier. It becomes an unrepeatable event or a quirk. The information is not absorbed or learned. And third, largely due to the first two effects, paradigms maintain the status quo. In other words, they provide protection. Obviously, change is almost never accepted because it deviates from the norm. The norm, whether for good or not, is comfortable and known. Therefore, paradigms become very valuable tools for protecting the norm or status quo.

Paradigms:

▸ Mask Recognition
▸ Limit Learning
▸ Maintain the Status Quo

That's why paradigms are important in our search for total quality. You have them and so do I. This material is being presented from my paradigm and it is probably very different from yours. The possible outcomes are to not see the information at all or limit what is learned and protect the status quo. Recognizing that these effects come from paradigms allows a more open–minded reading and, thereby, a more open learning experience.

PARADIGM SHIFTS

What happens when paradigms change? Joel Barker used many examples of paradigm changes or shifts. In those examples, he talked about the 1960's American mind-set about the phrase, "Made in Japan." Then, these words were synonymous with cheap, imitation, low quality, and easy to break. Something happened over the next twenty years, though—the paradigm shifted. Under the new paradigm, these same words, "Made in Japan" mean expensive, original and on the cutting edge, very high quality and durable. The paradigm had changed 180 degrees.

From this and many more examples, we have learned three important aspects of paradigm shifts. First, new ways of thinking come from the edge of the existing paradigm. So the new paradigms don't differ a little, they differ a lot! Ideas that differ only a little bit from what we expect are much easier to see and accept. These require only a slight change in our paradigms. Those changes that come from the edge are the ideas that are difficult to recognize, accept, or even question. Barker uses the illustration of the Swiss concept of watch movement. The Swiss had monopolized the world of fine watches. From their own researchers came the idea of quartz action in watches. Because it had few gears and no mainspring, it came from the edge of the Swiss Watch Paradigm. So, it was rejected. The result of their failure to see this important paradigm shift and, thus, shift their own paradigm was huge. Analyze the result for yourself. Today, what country has a monopoly on fine watches? Japan. They saw the new idea and recognized that it represented a paradigm shift. Because they did not share the Swiss Watch Paradigm, they took full advantage of the idea.

Second, paradigm shifts are not easily seen. Which idea from the "edge" represents a shift in the paradigm? Which one does not? Just because an idea comes from the edge of what we expect doesn't guarantee its success. Look at the Edsel!

The third aspect of paradigm shifts is that they are seldom discovered by accident. They must be actively sought. Sure, there is the story about the school teacher who "out of the blue" bought a bunch of stock in a little known company named International Business Machines. She retired rich. Examples like this are rare. People, companies, industries, and organizations that sit back waiting for a beam of light or voices from above announcing a shift in a paradigm will *not* succeed. Only those actively seeking the information that may cause a paradigm shift will be on the cutting edge of the movement and be able to both recognize it and take advantage of it. This is one of the main reasons that the Japanese have been so successful. They spend a great deal of time searching for new ideas so that they can take advantage of them.

Paradigm Shifts:

- ► Come from the "edge"
- ► Are NOT easily seen
- ► Are seldom discovered by accident

By knowing what went on in the past and why, you can avoid many problems in the future. History also helps you see paradigm shifts *before* they can hurt you. Americans tend not to be historians and, thus, learn little from history. If it isn't invented in America or considered to be new wave, Americans have little interest in it. Americans also have a history of not seeing important paradigm shifts until they are obvious. Americans simply aren't attentive to them. Total quality is an excellent example.

A poignant example of Americans not seeing important paradigm shifts happened more than fifty years ago. It involved a paradigm shift that was so powerful that it could have completely changed American culture and society. America's leaders were warned about the paradigm shift and they all saw the evidence that pointed to it. But, still they didn't see it coming. This paradigm shift involved priorities and strategies of warfare and big boats, specifically navy ships.

In the late 1930s and into the 1940s, the world was in real conflict. Both Europe and southeast Asia were fighting for their existence. America was helping a little, but it was mostly sitting back protecting its own interests. The American navy had an absolute paradigm concerning ships and hadn't noticed that the paradigm had changed and changed drastically. Thus, America was made vulnerable to disaster. Through direct action by another country, a stroke of luck, and some bad timing, the leaders of America were forced to change their paradigm. If they hadn't awakened and embraced the new paradigm, Americans could be speaking a different language today.

What was America's paradigm and how had it shifted? The American paradigm said that the strength of any navy was defined by the number of battleships that could be floated. Battleships were the strength and the most prized possession of America's navy. The Japanese, however, had accepted a different paradigm. They believed that the aircraft carrier was the backbone of their navy's strength. Because of that difference in paradigms, America all but ignored a floating armada of Japanese ships that left Japan for the Pacific in the fall of 1941. Luckily, the navy had sent an unknown strength—the new paradigm, the American navy carriers—towards Wake Island. America's pride and joy, its battleships, the out-of-date paradigm, were anchored safely at Pearl Harbor. The rest everyone knows.

The Japanese caught the American navy, air, and ground forces with their "paradigm" pants down. They pounded the American ground-based aircraft and sunk all battleships. In two waves, Pearl Harbor was left billowing in smoke—a burning, twisted wreckage. More than 2,000 American military personnel died because of America's out-of-date paradigm. The sad point is that America's leaders were warned by one of their best, Admiral Halsey. He saw the new paradigm. He had told his leaders about it. At the time of the attack on Pearl Harbor, he was steaming toward Wake Island in a ship that represented the new paradigm.

The Japanese didn't totally destroy America's navy. They didn't attack our submarines at Pearl Harbor and the American carriers weren't there to destroy. It was bad timing for Japan, good timing for America. The Japanese also didn't anticipate one of America's strengths; they did not recognize that once faced with an unavoidable, undeniable paradigm shift, Americans move quickly and with full determination. American carriers and carrier-based aircraft made the difference in turning the war effort at Midway. Americans not only learned the new paradigm, they played it better. I've got a paradigm question for you. Looking at total quality, do you hear the engines of in-coming aircraft?

Business provides many examples of failure to see a paradigm shift. One comes from IBM. During the technological revolution that caused vast changes in the computer market, IBM quit seeking a shift in paradigm. Instead, it became extremely protectionistic of the "Big Blue" line. IBM's paradigm assumed that there would always be a viable market for mainframe computers and that the computer world would conform to IBM because it was *the* industry leader. IBM believed its own paradigm so earnestly that it stopped selling its very successful small computer to mainframe customers. Because it stopped looking for the paradigm shift, IBM found itself in serious trouble. Record losses were reported, and the giant corporation was forced to reorganize. Thousands of people lost their jobs. Investors lost millions. Like the Swiss, IBM didn't recognize that paradigm shifts are not respecters of any past accomplishments or status. Paradigm shifts can also *not* be stopped, no matter how big or important the company. They simply happen, with or without the giants.

WHAT ARE YOUR PARADIGMS?

So, paradigms are really important to be aware of and respect. For our purposes, let's bring the discussion a little closer to home. What are your paradigms concerning total quality and, more appropriately, concerning safety and health programs? For example, do you believe:

- Safety inspections are very valuable because they identify unsafe conditions.
- "Unsafe action" is another way of saying "safety violation."
- The role of safety is to police safety violations.
- Establishing safety records, like going 1,000,000 workhours without a lost-time injury, are good indicators of a quality safety program.
- A strong, well-staffed safety department is the key to a safe workplace.
- The safety group should spend the majority of its time inspecting the workplace and investigating injuries and incidents.
- One main function of safety is to assure follow-up and/or correction of identified safety and health hazards.
- Safety and/or health is *MY* turf!
- Safety's main purpose is to protect workers.
- There will always be conflicts and competition between safety, quality, and production.
- Safety is Number 1!
- The key to a good safety and health program is to have support from upper management.
- The role of safety is to assure compliance.
- Some employees are just accident prone.
- There is no way that a safety group can make a facility safe and maintain compliance.
- Safety is a staff function.
- This total quality thing is diverting attention away from safety.
- A stronger OSHA and MSHA will improve worker safety and health.

- ▸ The key to making management accountable for the safety and health of workers is to enforce criminal penalties.
- ▸ We need to become as powerful as those "Quality" folks.

What kind of paradigms do you bring with you as read this book? More important, how are your paradigms going to affect your ability to recognize new concepts, test them thoroughly, and, if appropriate for your application, adopt them?

We all have our own safety paradigms. It's important to remember that we were not born with our paradigms. They are not life-dependent nor divinely inspired. And if we change a paradigm, life doesn't change; only our perspective or viewpoint changes.

My son came home from school one day saying, "My English teacher is a real idiot." "Why do you think so?" I asked. "She can't teach," was his only reply. Knowing that report cards were coming shortly, I asked, "What grade are you getting in her class?" "I'll be lucky if I get a D," was his frustrated reply. What was the real problem? Was his teacher a rotten teacher? No, I don't think so. Was my son a poor student? No, his GPA was just short of 4.0. So, what was the problem? My son's paradigm about his teacher was the problem. According to his paradigm, she could not teach and, thus, he could not learn from her and earn a higher grade. Working with him to change his paradigm quickly reversed the problem. His reward for changing his paradigm was an "A" in her class. Paradigms are critical to learning!

What do your paradigms say "reality" is? As we continue our total quality journey, how will your paradigms affect your learning?

3

WHAT IS TOTAL QUALITY MADE UP OF?

Now our subject gets more interesting and a lot more detailed. Remember: total quality is not an object. It's a process. And describing the makeup of a process is much different from describing what composes an object, such as a bicycle. For example, a bicycle is made up of various parts—wheels, frame, chain, spokes, etc. So, an object is composed of smaller objects or parts. Unlike a bicycle, a process is not tangible. A process isn't made up of objects. It is built upon concepts and has various components or tools that are used. Concepts can be thought of as the foundations upon which a process is built. Components, like tools, are used to make a process work.

So, how can we identify what a process is made up of? Seems like a contradiction. Making an analogy to another process will help explain.

One of the processes that Americans revere is that process that has long defined our country—freedom. Freedom is very familiar. So, it should serve as a good analogy. Is freedom made up of any objects or parts? It really doesn't have any parts or pieces. Freedom has some very common concepts and components though. Those would include self-determination, certain rights such as ownership and voting, and some responsibilities like following society's rules. None are tangible, but they are definable. So, freedom *does* have some common concepts and components. And we can use those concepts and components to describe what the process of freedom is all about. This is exactly what needs to be done for total quality.

THE CONCEPTS AND COMPONENTS OF TOTAL QUALITY

Studying successful total quality programs, we find six common, specific program concepts and components. Specific applications or specialties may not require all of them. The six are: total quality programs have common focuses, use dynamic people concepts, employ strategic planning, set objectives and measure progress, benchmark and evaluate, and believe in continuous improvement.

First, total quality programs have *common focuses*. What do they focus on? Total quality programs focus on customers, wastes, time, and excellence. Focus is a measure of intensity; it isn't casual and it isn't an awareness or knowledge of something. It's an intense concentration and requires extreme, mind-and-body dedication and commitment. Look at any NFL football team at the beginning of the season. What is everyone on the team, on the sidelines, in the front office, and in the stands concentrating on? The Super Bowl! All are convinced that with a dedicated, full-team effort, they can and will make it to the Super Bowl. And given that chance, they will win it all. It isn't a maybe. Like Mike Ditka recalled from the Chicago Bear's Super Bowl year in 1985, from the first moment in training camp, everyone on the team *knew* they were going to win the Super Bowl. That's focus!

In the focus area of customers, it's more than merely knowing who your customers are. It is a total focus on them, what they need and what they expect. Customer focus is one of the things that separates business superstars from also-rans. McDonald's and Disneyland are excellent examples of customer focus. Walt Disney put it this way, "We want to do our job so well that they will want to come again, and bring a friend." That's customer focus. Customer focus goes much deeper than just external customers. Each organization, program, business, etc., also has internal customers. Internal customers would include management, employees, departments that take semi-finished goods from your department, etc. In a total quality program, internal customers are no less important that external ones.

Focusing on wastes requires the same level of concentration and dedication. Don't just focus on the reductionistic definition of the term. Wastes aren't only what is thrown away or disposed of. Wastes include anything that is inefficient, wastes resources, or generates waste. Examples would include product rework, inventory, disposal costs, too much labor, injuries, illnesses, having to walk too far to do a job, etc. Wastes are inefficient and drive costs up. Good total quality programs have to focus on wastes and their elimination.

Successful total quality programs focus on time. Time applies to wastes, but because it is so important, it is given a focus of its own. The important time element is called cycle time. Cycle time is the time from the start of an activity to completion of the activity. It can be all-inclusive, like the time it takes to make a product from raw material until it is finished and shipped to the customer. It can also be a portion of the process such as the time it takes to complete a task.

Why is cycle time so important? Because the longer it takes to do something, the more it costs. There are two important areas of wasted time: Q-time and set-up time. Q-time is nonproductive time: time spent waiting for parts, waiting for equipment availability, waiting for inspection, waiting for paper or directions, waiting for approvals, waiting for decisions, etc. Also important is set-up time. Unlike Q-time, people remain busy, either setting up or changing over some operation or piece of equipment. Changing machine dies and setting up a machine for precision work are good examples of set-up time. Q-time and set-up time are two main extenders of cycle time and both affect product and operation costs.

Successful total quality programs have to focus on excellence. Excellence is the outcome of the total quality process. Anything short of excellent is second-rate. Excellence can only be determined by your customers and only by comparison. As Lee Iacocca said, "We want to be the best. What else is there?" Excellence is a continual pursuit. Like the other focus areas of successful total quality programs, excellence requires total concentration and commitment. No compromises. Nothing else will do.

Second, total quality programs have *dynamic people concepts*. People are one of the most important resources to a successful total quality program. The history of management and resulting management theories show that the importance of people has cycled up and down. Theory X management, for example, considered people as necessary resources, much like raw materials. Employees were there to do work, and work was to make a product or perform a service, nothing more, nothing less. A good worker was a productive worker, a happy employee, and a secure employee. At least, he or she was secure as long as he or she performed at the expected productivity and quality level, and provided that the managers ran the business well. It was a "your world—my world" concept. Management ran the business and made the decisions. Workers operated machines, inspected product, shipped product, unloaded raw materials, etc. They were not included in the decision-making process.

Theory Y of management came along and, in the companies that used it, the importance of people leapt to the opposite end of the continuum. Under Theory Y, people were everything! Managers became known as "huggers and kissers," catering to employee needs, keeping them happy. In Theory Y, employees had all the answers. Management was there to coordinate and serve. Businesses have been successful operating under both Theory X and Y. In a total quality environment, neither works.

Theory Z management, also commonly known as Japanese Management (although it isn't totally), is better described as team management or participative management. Originally described in Japanese industry, this type of management used team dynamics to solve problems and make decisions. It has been very successful. Theory Z management, however, devalues the individual. Teams are the only recognized entity. To "belong" you have to be part of a team. Recognition and rewards are focused on the team performance and successes. Americans, who have long admired individual effort, have had trouble adapting to this management style, primarily due to this individual devaluing. America's entrepreneurial society holds strongly that individuals who risk and succeed get ahead. Successful total quality programs need to blend the synergy and dynamics available within the Theory Z school of thought with a modification of Theory X management that is now called leadership and with the entrepreneurial concept of individuality. Total quality uses people concepts, group and individual, as a dynamic tool.

Within these people concepts utilized by total quality programs, three notable terms stand out: participation, empowerment, and ownership. Participation is a deep concept that must be fostered, nurtured and supported to be sustained. It not only speaks of involvement at all levels of an organization, top to bottom, but also includes involvement in most of the areas that affect the organization. For participation to occur, a number of interrelated needs must be met. These needs include opportunity, knowledge and training, time, commitment, resources, support, encouragement, and responsiveness.

Empowerment, to a great extent, is a measure of the autonomy awarded to an employee or group of employees. Autonomy is a function of trust and challenge. In his book, *The Game of Work*, Chuck Coonradt set a special, large area in the middle of his "Field of Play," which was his way of describing an employee's work. This area defines the concept of empowerment very well. In this middle area, employees do their jobs, independent of being overly managed. Coonradt called this area the "G.O.M.B."—or Get Off My Back—Zone. That's what empowerment is: letting an employee do his or her job without over- or under-management and control. Peter Drucker in his seminal book, *Management*, wrote, "The purpose of an organization is to enable common men to do uncommon things." That's empowerment!

Ownership is accountability, recognition, and reward. Efforts together with success are recognized personally and publicly. Reward is very personal. Whatever rewards are used, they must be selected and made appropriate to the employee and the organization's culture.

Third, total quality programs use *strategic planning concepts* as a tool to measure their progress and plot their pathway to success. Strategic planning incorporates three interrelated levels: strategic, long-term, and short-term. Strategic planning is what has to be accomplished now and in the future to achieve set goals; strategic plans change as the organization makes progress and as standards change. Thus, strategic plans change a lot. Long-range plans are more concrete, are usually progressive, and are aimed at achieving your strategic plan. Their term is usually three to five years. Short-term plans are bite-sized pieces of the long-range plan; they are more detailed and usually cover a year of operation. The strategic planning process is critical to a successful total quality program. Without it, any success happens by chance, luck, or accident.

Fourth, total quality programs *set objectives and measure progress* toward them. Objectives not only complement the planning process, but identify the "have to's," the "need to's," and the "want to's" that should be accomplished. Objectives by themselves mean little to the success of an organization; objectives must be accompanied by measurement. Measurement allows everyone to know if progress is being made, if the path needs correction, and when the objective has been achieved. Objective setting is important, but measurement is critical.

The fifth component of the total quality process is *benchmarking and evaluations*. Benchmarking is an improvement tool which front-running organizations, companies, or programs use as models. Improvement objectives are based on reaching those models. Benchmarking also includes a process of measuring that improvement. Benchmarking is a very valuable tool that avoids, to a great extent, the need to "reinvent the wheel." Model programs and organizations usually invite benchmarking because it "shares the wealth." It's also a real ego boost.

Program evaluations are systematic tools for assessing program growth, areas where improvement can be made, or where programs are strong. Benchmarking without program evaluations is like icing without the cake. Through program evaluations, good programs can become great programs. Through benchmarking, great programs can become industry leaders.

The sixth total quality process concept is *continuous improvement*. This thought runs counter to traditional American industrial thought of "just throw money at it." Continuous improvement is not just an employee empowerment issue. It also focuses on small, low or no-cost improvements. Classical American industrial thought says that to improve a process, a new process, or assembly line, or automation is necessary. Those cost a lot and aren't always successful. Continuous improvement takes ideas from the "grass roots" level of an organization and implements those small improvements. It could be as simple as modifying a hand tool, elevating a work surface, increasing the lighting, moving a storage cabinet closer to a worker, or rearranging equipment to smooth out product flow.

First, by implementing small changes, major costs are controlled. Second, employees are involved in the improvement process. Third, little or no interruption occurs in production. Fourth, employee morale and enthusiasm are built. And fifth, major capital expenditures may be completely eliminated or changed so that they can be more effective. Chapter 16 is devoted to talking about continuous improvement. In a total quality organization, continuous improvement is that important.

Total Quality Programs:

- Have common focuses:
  (Customers, Wastes, Time, and Excellence)
- Have dynamic people concepts
  (Participation, Empowerment, and Ownership)
- Use strategic planning
- Set objectives and measure progress
- Use benchmarking and program evaluations
- Believe in continuous improvement

JUST-IN-TIME MANUFACTURING

The dynamic process of total quality has multiple origins and numerous influences in its development. Current total quality tools include a very significant one, "Just-in-Time,"[1] or JIT. The focuses and concepts of JIT make it inseparable from total quality. JIT is mostly of Japanese origin. JIT is a cost-reduction program that focuses on work efficiency and waste elimination. Does that sound familiar? Its basic premise is that there are better, easier, and less costly ways to do business. With JIT, most companies do more with fewer people and significantly reduce their operating costs.

---

[1]D. J. Lu, trans., Japan Management Association, ed., *Kanban, Just-In-Time at Toyota*, (Cambridge, Mass.: Productivity Press, 1985.)

Just-in-Time manufacturing has eight elements: pull-system, multifunctional workers, level production, continuous flow, product quality control, set-up elimination, machine up-time, and housekeeping and safety. Let's discuss each of these briefly.

Traditional American manufacturing uses what is called a push-system. Production operator #1 makes widgets, usually on a piece-rate basis. So, to operator #1, more widgets made translates into more wages paid. After the widgets are made, they are sent to operator #2 who builds them into widget assemblies. If operator #2 can produce at the same rate as operator #1, all is well. However, if he or she can't, the push-system begins to cause problems. For example, if operator #2 makes the assemblies faster than operator #1 provides the basic widget, operator #2 spends a lot of time waiting, not earning money. Obviously, a production inefficiency results as well as a serious personnel problem. If, however, operator #2 produces at a rate that is slower than #1, operator #1 will not decrease his or her production *and* pay, so widgets begin to stack. This results in a lot of in-process inventory that costs money.

Pull-systems work in reverse. Operator #1 doesn't make a widget until the one he or she made last is taken by operator #2. In such a system, you are forced to know a lot more about the process itself and how long it takes to do each specific step so efficiency can be built into the system. Pull-systems save a great deal in inventory costs and greatly reduce error rates.

The second element of JIT is the *multifunctional worker*. In JIT you break down the classical job description barriers that specifically define a worker's role, e.g., lathe operator, machinist, electrician, material handler, crane operator, etc. Multifunctionality is simple. If workers are proficient at several jobs and tasks, they can be used more efficiently. They can be used *where* there is need, not *if* there is need.

The third element is *level production*. Level production takes the peaks and valleys out of the production schedule, e.g., January production = 1,103, February production = 379, and March production = 1,890. To illustrate the benefits of level production, imagine the freeways and interstate highways in your area being level-loaded. There would be no commuter traffic jams in the morning and afternoon because traffic would be spread out over the entire 24-hour day. Every hour of the day, the traffic is the same. Think of the advantages to travel at *anytime* of the day.

Fourth, JIT uses *continuous flow*. Repeatedly starting up a process and shutting it down is very inefficient and wastes time and money. Continuous flow maintains a constant level of operation within the level production requirements. Product doesn't sit and wait for use or continued processing. In-process inventories are thereby reduced.

Product quality control, or QC, is the fifth JIT element. Here is a more subtle difference. In JIT the classical "Quality Control" department function is challenged. It just makes better sense to move QC onto the shop floor and build inspection skills into the operators. It makes them more aware of quality problems, more efficient at identifying the causes of those problems, and increases the amount of QC inspections that a product receives.

Sixth, JIT seeks to *eliminate set-up*, or at least make it more efficient. One of the acronyms used in JIT best describes this element, SMED. SMED stands for Single Minute Exchange of Dies. The focus here is to make die changes in machinery as efficient as possible, thereby reducing major wastes, equipment down-time, and nonproductive employees.

The seventh element of JIT is *machine up-time*. This concept also runs counter to classical American industrial thought. Historically, American industry has focused on decreasing down-time, the exact opposite of up-time. In decreasing down-time, the focus is on trouble calls and maintenance calls. In increasing up-time, the focus is on time the machine is being used in production. As an analogy, think of tracking down-time as how often your car is in the shop for repairs and how much it costs to get it fixed—interesting data but not very useful if you still drive the same distance and the same number of days. Your down-time may decrease but your utilization of that resource is the same. What value did that give you? Zero. If, however, you focused on increasing the distance, or days you use your car or car pool as measurements, then you are receiving added value from your focus, better use of your car.

The last element of JIT is *housekeeping and safety*. The rationale is three-fold. First, safety and good housekeeping go hand-in-hand. Second, as Taiichi Ohno, the father of Just-in-Time manufacturing said, "You can't make a quality product in a pigsty." And third, as we know, poor safety is a waste and costs money.

The Eight Elements of Just in Time:

1. Pull System
2. Multifunctional Worker
3. Level Production
4. Continuous Flow
5. Product Quality Control
6. Eliminate Set-Up
7. Machine Up-Time
8. Housekeeping and Safety

With the eight elements of JIT, there are also seven deadly wastes. These deadly wastes become the focus for improvement. They are: over-producing, time on-hand waiting, transporting, the process itself, unnecessary stock on hand, unnecessary motion, and producing defective goods. We'll revisit these seven deadly wastes later in this chapter.

The Seven Deadly Wastes of Just-in-Time:

1. Over-Producing
2. Time On-Hand Waiting
3. Transporting
4. The Process Itself
5. Unnecessary Stock On Hand
6. Unnecessary Motion
7. Producing Defective Goods

JIT: ADDITIONAL CONCEPTS

Other tools and concepts used within JIT are useful in total quality. Those include line stop, work cells, poke-yoke, kanban, and product flow. **Line stop** is simple. When a defect is identified, the production line stops and all attention is directed to finding the cause and correcting it. This avoids making more defective products and increases the problem-solving resources at the failure point. This isn't a new concept in safety. In fact, safety has used line stop for years—it's called an accident investigation. This, however, is not the best use of line stop. How often is line stop used when a hazard is identified or an incident without injury or property damage occurs? Makes you stop and think.

Work cells are an attempt to move equipment, tools, product flow, etc. into an as efficient as possible work space. This allows for better use of equipment and manpower, and decreases wastes such as time waiting and employees transporting materials or walking to and from different machines. To a large extent in the field of safety and health, it's ergonomics.

Poke-yoke is a Japanese term that means to fix something so it will not happen again. It uses two separate actions: 1) find the true cause of the event, and 2) fix it. The problem-solving part can be very complex. There are some skills and training required to accomplish this part of the task. The fixing part can also be complex. Poke-yoke accepts no "band-aids." If a fix is in order, it must prevent recurrence of the event or it is no good. The poke-yoke concept is simple, however. If an event is fixed once, it will not recur. As events are poke-yoked, the problems melt away one by one and future problems happen less. It's kind of like a half-life calculation in which the number of events, as they are poke-yoked, get fewer and fewer and fewer. So, the production, quality, and/or safety improves and improves and improves.

The **kanban** concept complements the pull-system. Originally, a kanban was a card that announced a need to start production. More broadly, it is any mechanism that does so. It is an important part of the pull-system. In the JIT system, whether or not a department or operator begins production depends on if there is a kanban, or need. Let's return to our simple description from the pull-system. Operator #1 will not begin to make a widget until operator #2 has a need. That need is conveyed by use of a kanban. Kanbans can also be layout places on the shop floor, in-baskets, or prompts on a computer screen. In reality, a kanban can be anything that communicates a need and triggers starting a product or service, like safety and health.

In JIT, there is almost a religious passion for **product flow** or, more specifically, to streamlining product flow and making it more efficient. For example, a product flow that zigzags across a shop floor or moves from one building to another and then back, is not only a hazard, it's inefficient. Likewise, focusing on product flow helps identify those steps that do not "add value" and, thus, can be dropped. Product flow is a very important focus in the JIT and a total quality system.

With the continual evolution of total quality, the lines separating processes such as classical total quality and JIT have disappeared. The processes overlap and act to synergize the entire total quality process. Remember that total quality is always changing. In that light, there are three more total quality tools. These additional processes are: process mapping, CEDAC, and X-Matrix.

Process mapping[2] (Figure 1) is an analytical tool that breaks down a process, production or administrative, into *all* of its steps. It stresses Q-time, inefficiencies, and redundancies. Process mapping allows easy identification of those areas of the process that do not "add value" to the system so, like a line stop, attention can be directed to eliminating or minimizing the inefficiency.

---

[2]K. Ozeki and T. Asaka, *Handbook of Quality Tools* (Cambridge, Mass.: Productivity Press: 1990.)

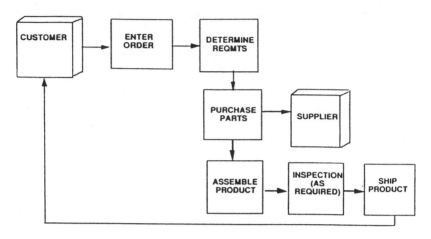

Figure 1: Process Map of a Simple Function

CEDAC[3] stands for Cause-and-Effect Diagram with the Addition of Cards. CEDAC is a problem-solving tool aimed at continuous improvement. It starts when a problem is identified. Working as individuals within a team problem-solving effort, each participant places cards on the CEDAC diagram (Figure 2). Each card contains one idea on how to solve the problem. There is no such thing as a good card or a poor one, a smart idea or a dumb one. The goal is to get as many ideas as possible, including ideas that are logical and those that come from the edge. Each idea is evaluated and discussed. One by one, the cards are removed (only by the person who placed the particular card there) until the one or ones that remain can't be easily discarded. Those are prioritized and implemented. The results of each are measured. CEDAC is an excellent tool for focusing a team effort on a particular problem and soliciting the maximum number of possible solutions and participation.

---

[3] CEDAC is a registered service mark of Productivity, Inc. R. Fukuda, *CEDAC: A Tool for Continuous Systematic Improvement* (Cambridge, Mass.: Productivity Press, 1986.)

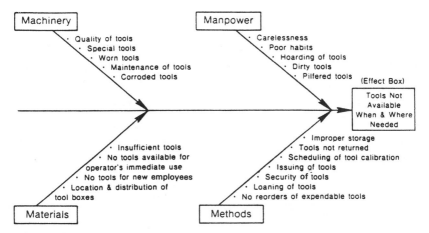

Figure 2: A CEDAC Cause and Effect Diagram

X Matrix[4] is a process that builds team efforts and communication. It is especially good for projects or activities that require support and action from many different departments or functions that do not have common reporting structures. A worksheet (Figure 3) is used for easily identifying the steps that require support or action by other groups or departments. X Matrix is extremely useful for service- or staff-oriented functions such as safety or engineering. It serves not only to identify needed support, but also to get that support for a common project or goal.

IS TOTAL QUALITY APPLICABLE TO SAFETY AND HEALTH?

It's natural at this point to ask, "That's an interesting brief primer on total quality, but what does it really mean to me in my safety and health program?" Good question. As stated earlier, there are two possible positive results from reading this book: you can put the total quality

---

[4]S. Mizuno, *Management for Quality Improvement* (Cambridge, Mass.: Productivity Press, 1988.)

process into your safety and health program, or you can more effectively put your safety and health program into a total quality organization. If the second is your intended goal, knowledge of the total quality process and tools will help. In that type of application, it isn't "them or us." It's safety as a part of the total quality process that everyone is working toward. If, however, the first possible positive result is your goal, let's look at how the classical manufacturing total quality process can be converted to a service application.

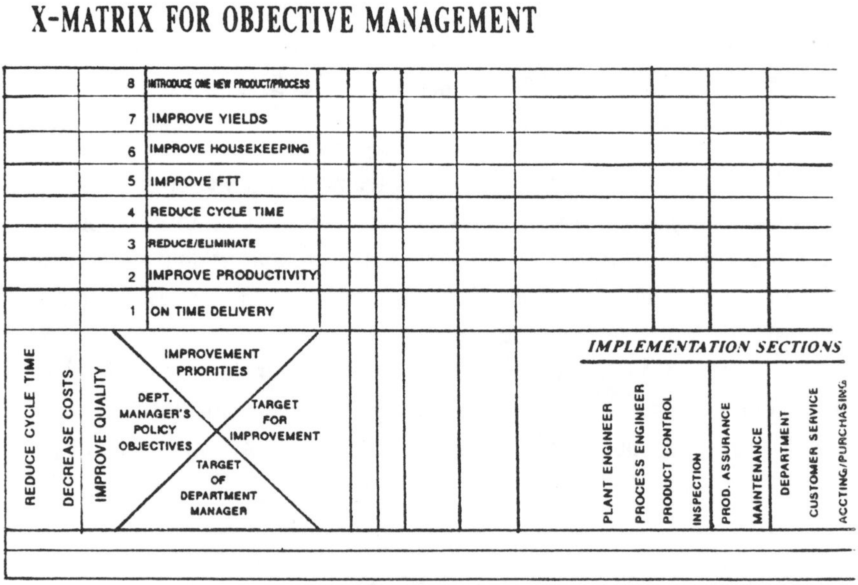

X-MATRIX FOR OBJECTIVE MANAGEMENT

	8	INTRODUCE ONE NEW PRODUCT/PROCESS															
	7	IMPROVE YIELDS															
	6	IMPROVE HOUSEKEEPING															
	5	IMPROVE FTT															
	4	REDUCE CYCLE TIME															
	3	REDUCE/ELIMINATE															
	2	IMPROVE PRODUCTIVITY															
	1	ON TIME DELIVERY															

REDUCE CYCLE TIME — DECREASE COSTS — IMPROVE QUALITY

IMPROVEMENT PRIORITIES

DEPT. MANAGER'S POLICY OBJECTIVES — TARGET FOR IMPROVEMENT

TARGET OF DEPARTMENT MANAGER

IMPLEMENTATION SECTIONS

PLANT ENGINEER — PROCESS ENGINEER — PRODUCT CONTROL — INSPECTION — PROD. ASSURANCE — MAINTENANCE — DEPARTMENT — CUSTOMER SERVICE — ACCTING/PURCHASING

Figure 3: An X-Matrix Sheet

The perceived differences in total quality application, manufacturing versus service, exist *only* in one's paradigms! Because of this, it is important first, to recognize the fact, second, to open your mind to a new way of thinking, and third, to actively challenge your paradigms.

Total quality programs have common focuses: customers, wastes, time, and excellence. Traditional industrial culture has been spotty and mixed on knowing who the customers are and how important they are. As an example, during the 1960s and into the 1970s, the American automobile manufacturing industry had such strong beliefs about this subject that their products were affected. They thought customers would buy what Detroit manufactured. It was a "tell-you" customer focus, not a "listen to you" customer focus. Today, there is much more listening than telling. Why the change? Look at the market share and profit trend of American automobile manufacturers over the past twenty years.

In today's total quality organizations, the customer has wide definition, but the importance is always "Number 1." Focusing on what the customer wants, needs, and expects is a critical part of business. The definition of customer has been expanded, however. Today, customer means not only those who purchase products, but also employees, different departments within the organization, and line management.

How is this aspect of total quality customer focuses applicable to service and staff functions like safety and health? Who are your customers and how important are they to your program? Historically, if you'd asked a safety or health practitioner, "Who's your customer?," you would have heard, "The employee." In a total quality program, that definition is expanded. Employees are still one of the highest priority customers of any safety and health program. Today's definition, however, includes management at all levels, other staff functions, matrix reporting positions, and those who have chosen you for benchmarking. What is the ranking of priorities? Each will have its time at the top of your priority grid. Overall, however, each has pretty much the same priority—high.

Do you know what your customers want or expect from your service? Do you regularly ask them what they want or expect or do you just assume that you know?

American industries' focus on wastes is new. One doesn't have to go back very far to remember a near-sterile Lake Erie, smokestacks belching black smoke, and hazardous waste disposal via barges and the open sea. However, in total quality businesses today, the definition of wastes has been expanded. It includes wasted inventory, production, re-work, manpower, inflated prices for materials or services, storage, too much material transportation, injuries and illnesses, inefficient manufacturing processes, production stops, imperfections in material quality, returned product, product liability issues, loss of employees, to name just a few.

Focusing on waste allows wasted effort and costs to be eliminated or minimized, thus allowing a business to be more productive and make more profits.

Is this focus on wastes different in a safety and health program? No. Focusing on wasted effort, expenses, etc., allows the program to be more productive and to avoid costs—ultimately improving the business's profit margin. It's important to note that aside from consultation, safety and health programs generate no revenue to the business. Therefore, cost efficiency is key. What kinds of wastes can a safety and health program focus on? A short list would include the wasted effort in doing a routine task such as dispensing safety glasses or approving chemical purchases. It would include wasted costs of safety equipment such as gloves or respirators, wasted time responding to recurring problems such as accidents and communication difficulties. It would also include the wasted effort in unproductively chasing or finding anything. Get the idea? Obviously, an analysis begins with knowing where your time goes. Which, naturally, deserves to be the next waste in our discussion.

Time is probably the greatest resource that is wasted. After all, what's your number one cost of doing business? People. Inefficient use of people costs money, lots of money. That's why total quality organizations focus on time. How much time does it take to make a product, do a task, or perform a function? This can be determined by breaking the time into both value-adding time—that time when something is actually being done to further a product's manufacture—and nonvalue-adding time—that time spent waiting, setting up equipment, transporting material, putting parts into and taking them out of stock, getting needed tools or equipment, starting up or shutting down equipment, etc. Eliminating or minimizing the nonvalue-adding time increases the efficiency of this most costly resource, people.

Is focusing on time different in a safety and health application? Not at all. People are still our most valuable resource. Anyone who practices safety and health has a limited amount of time to work with. Total quality safety programs focus on wasted time in both routine and nonroutine functions. How long does it take to do a safety inspection, complete and review an accident report, approve a new chemical product, measure an employee's exposure, do a physical examination, do a monthly report, have a problem-solving meeting, file records and reports, ready a shipment of hazardous waste, provide safety training, etc.? Starting with the tasks that happen most frequently or take the most time, and analyzing them for steps and then for value-adding and nonvalue-adding time allows a function or task to be streamlined and made more efficient. Process mapping is an excellent tool here. Efficiency allows everyone to spend their time on more worthwhile activities, such as proactive safety and health functions.

For too long American industry accepted the paradigm that products were supposed to break and, thereby, require repair. It wasn't until more forward-thinking players such as the Germans and the Japanese came along that our thinking changed. And these established players are continuing to improve as are new players entering the game. Excellence is being defined and redefined each day. If excellence is the only road that assures tomorrow, this focus on excellence in a total quality business is a matter of survival. Products must be made better, with higher quality at competitive prices, last longer, provide more no-cost features, and require less repair. The customer expects it more today than ever before and will expect it even more tomorrow.

How about excellence in safety and health? For too long we've been focused on the wrong area when it comes to excellence. We've focused on our injury rates, lost work days, and workers' compensation costs. These are the wrong indicators of excellence because they are influenced by too many things outside the safety and health program. For example, using injury rates as a measure of excellence will place good and bad programs on an equal footing, even if different injury severities are involved or games are being played with the numbers. If workers' compensation costs are used as an indicator of excellence, a program can get significantly worse just by a new doctor coming into the area who is ignorant of occupational needs and practice or workers' compensation rules, or by a new surgical procedure becoming vogue (e.g., Carpal Tunnel Syndrome surgery). These traditional indicators are very poor measures of the quality of a safety and health program. They cast too many "shadows" that draw attention away from what makes a program excellent. What should a safety and health program focus on to move toward excellence? It should use measures like customer surveys, multifaceted program evaluations, and benchmarking. These are true indicators.

Total quality programs use dynamic people concepts: participation, empowerment, and ownership. American industry has made a lot of progress in people concepts over the past ten years or so. There are still *wide* ranges in the application of these people concepts, however. They range from not-at-all to fully ingrained in the organization's culture. Using these people concepts, long-held definitions of worker and management roles are being challenged and discarded. Long-standing walls between management and labor are being torn down. Communication pathways are being redefined. Critical aspects of these people concepts include trust, active communication, autonomy, involvement, sharing, accepting responsibilities, supporting other team members, and working together.

One valuable and widely used people concept is team-building. Away from the traditional definition that was synonymous with working team or work crew, today's team transcends the horizontal and vertical aspects of an organization. Team members come from different depart-ments, trades, staff units, different levels of management, and from outside the organization. The teams themselves are dynamic. They come together to solve a problem or accomplish a task. Once completed, the team dissolves only to have the members become involved in new teams and new missions.

There's a common but destructive thought in many safety programs. The thought goes like this, "If I, as the safety and health professional, do all the work and have all the knowledge, it is my program and I'm irreplaceable." It is called "turf." In a total quality safety and health program, the use of dynamic people concepts challenges the basic premises of "turf." Educating and including other people in all safety and health activities has both positive and negative aspects. If one is insecure about his or her abilities and engages in "turf," this letting-go may be too threatening. The safety and health professional would view this as a significant negative. If, however, he or she is secure and wants to be able to accomplish more, team-building has nothing but positives. Through using people concepts such as team participation, line responsibilities can be more firmly entrenched and reinforced. Additionally, the number of projects becomes disassociated from the amount of available time. The safety and health knowledge level increases significantly in the organization. And, the number of safety and health disciples in the workplace can rise. Because this is such an area of opportunity, Chapter 7 has been devoted to addressing people concepts in safety and health programs.

Total quality programs use strategic planning. Successful companies know where they are going and plan how to get there. How many safety and health programs operate by strategic planning? From my observations I would guess that the percentage of those that do would be considerably below 5 percent. Historically, safety programs, especially, could best be described as "Management by Fire Drill." They are so busy "putting out the fires" that they never get into planning what they *should* be doing. Total quality safety and health program *must* be managed by a strategic planning process. Preferably, the strategic plan is derived by a team and complements the organization's plan. To a total quality safety and health program, strategic planning is not a choice, it's a necessity. Chapter 11 talks about the process of strategic planning.

Total quality programs set objectives and measure progress toward them. This is another obvious statement about successful companies. Objective setting and measurement are ways of life for them. How about most safety and health programs? Do they use objective setting and measurement? Taking away the traditional measurement approaches that don't work, injury rates, lost workdays and compensation costs, I would guess that, again, less than 5 percent do. Objective setting in most safety and health programs is a day planner. Measurement of accomplishments is also almost totally absent in the safety and health fields. Setting objectives and measuring progress is critical to a total quality safety and health program. Chapter 11 also discusses objective setting and measurement.

Total quality programs use benchmarking and program evaluations. Most businesses neither benchmark nor do self-evaluations. Successful total quality companies always do. Why is benchmarking so important? First, it's hard to know how to improve without a model. Benchmarking can provide that model. Second, inventing everything is very costly. Benchmarking is a learning tool that avoids the need to "reinvent the wheel" at each company. Equally important, it also can help you learn from the mistakes others have made.

Is benchmarking important to a total quality safety and health program? If the program is focused on excellence it is. Again, how many safety and health programs today use benchmarking? Almost none. Why? There are two reasons. First, successful safety programs have historically been kept "close to the chest." There is a mystique about them, a mystique that seems to be driven by superstition. Second, pride (or false pride). To ask someone else why their program was successful might seem to be an admission that the asker doesn't know what he or she was doing. In many total quality companies today, you are beginning to see staff functions such as Human Resources (HR) begin to benchmark across to HR programs at other companies. It hasn't trickled down to safety and health programs much, but it will.

Program evaluations, in many forms and different applications, have existed in safety and health programs for many years. In countless different formats and levels of detail, program evaluations have existed to meet the perceived needs of the safety and health program itself. Those who foolishly feel that minimal self-evaluation is necessary use minimal or no evaluations. Performing programs that find greater value in evaluation use detailed, team-oriented self-evaluations at regular intervals. Some use program evaluations that are free-form, asking basic questions to guide group discussion. Some use extremely detailed program evaluations that are literally small books.

The purpose of program evaluations, of course, is to provide information for improving programs. This is a *very* fundamental concept of total quality.

Benchmarking is an optional exercise once a highly evolved or improved status has been reached. Some programs use benchmarking to reach a higher level. For any competent safety and health program, especially one aspiring to total quality, program evaluations are *not* optional. They are too fundamental to total quality. If benchmarking is not pursued, that's fine. If, however, program evaluations are unthinkable, quit thinking about total quality! To add more detail and discussion, Chapter 15 is included because benchmarking and evaluations are so important to total quality safety and health programs.

Total quality companies believe in continuous improvement. To most safety and health programs today, however, this is a foreign term and a foreign concept. The concepts and tools of continuous improvement are just as applicable to safety and health programs as they are to manufacturing. Because we would be able to do little more than add confusion here with a short discussion of continuous improvement, Chapter 16 is dedicated to it.

JIT IN SAFETY AND HEALTH PROGRAMS

Are the total quality tools of Just-in-Time applicable to safety and health programs? You bet they are! Here are examples of the Eight Elements of JIT that can be applied to total quality safety and health programs.

1. Pull System:
    ▸ Fire extinguisher service based on need, not on schedule.
    ▸ Waste pick-ups based on need, not schedule.
    ▸ Respirator cleaning and repair based on need, not schedule.
    ▸ Safety inspection frequency based on employee safety surveys and other means, not by schedule.
    ▸ Safety training focus and frequency based on random retention testing, not on preset schedules.

2. Multifunctional Worker:
    ▸ Cross-trained safety, environmental, industrial hygienists, etc.
    ▸ Multifunctional safety/health/other team audits.
    ▸ Hourly and supervisory personnel included in inspections, audits, training, and investigations.
    ▸ Detailed, ongoing training of line management in safety, health, and environmental issues.

3. Level Production:
   ▸ Safety and health activities scheduled equally throughout the year.
   ▸ Safety and health programs managed by planning versus "fire drills."

4. Continuous Flow:
   ▸ Using kanbans to trigger safety functions so that more is accomplished each time the function is triggered.
   ▸ Streamlining safety and health approval/review programs to minimize delays and customer waiting.

5. Product Quality Control:
   ▸ Multiple oversight of safety functions.
   ▸ Use of professional teams to accomplish tasks and check quality.
   ▸ Use of control programs rather than oversight.
   ▸ Have department ownership in equipment inspections and function testing.

6. Eliminate Set-up:
   ▸ Streamline safety and health administrative duties to eliminate nonvalue-adding activities.
   ▸ Use process mapping for time-consuming or frequent duties.
   ▸ Install satellite emergency response equipment areas.

7. Machine Up-time:
   ▸ Use teams to make more time available.
   ▸ Assign multifunctional professionals to tasks by need rather than by specialty.
   ▸ Emphasize proactive safety programs.
   ▸ Use strategic planning to maximize use of time.

8. Housekeeping and Safety
  ► Housekeeping emphasized in area audits and inspections.
  ► Safety becomes a line responsibility.

Surely you could significantly add to this list of examples, and if you can think of some, you have already begun to shift your paradigm. Once that shift from the position of "not applicable" to "a valuable resource" has begun, the sky's the limit to your applications.

How about the seven deadly wastes? Are they applicable to safety and health programs? Yes! Here are some applications to safety and health.

1. Overproducing:
  ► Conducting multiple safety training sessions that, if better coordinated, would require significantly fewer sessions.
  ► Writing a 30-page monthly safety report that isn't read when a "bullet list" and graph would be better.
  ► Conducting monthly safety inspections when the skills and responsibility could easily be passed on to the departments.
  ► Having multiple, unnecessary meetings.

2. Time On-Hand Waiting:
  ► Waiting for accident reports or reviews.
  ► Waiting for late meeting attendees.
  ► Waiting for decisions or direction.
  ► Waiting for safety to complete a review or hazard analysis.

3. Transporting:
  ► Driving to and from isolated work areas or locations.
  ► Traveling to corporate meetings.
  ► Traveling to and from unnecessary meetings
  ► Taking industrial hygiene equipment daily to an area to perform a multishift sampling.

4. The Process Itself:
   ▸ Beware of phrases that are used to describe or defend safety and health programs such as, "It's always been done that way," or "It's always worked best that way."

5. Unnecessary Stock On Hand:
   ▸ Too much safety manpower focused on a specialized component of the program.
   ▸ Warehouse safety stock that hasn't been turned over in more than a year.
   ▸ Specialized, seldom used safety supply inventory.
   ▸ "Just in case" stocking of safety equipment.

6. Unnecessary Motion:
   ▸ Observing industrial hygiene sampling for the entire sampling period.
   ▸ Having all emergency supplies centralized.
   ▸ Attending meetings when phone calls will do.
   ▸ Writing memos when a phone call will do.

7. Producing Defective Goods:
   ▸ Making recommendations without consulting the area workers.
   ▸ Assigning unrealistic completion goals.
   ▸ Taking too long to do reviews or give approvals.
   ▸ Making incorrect judgments from industrial hygiene sample results.

Total quality is a process for improvement. It is no more complex than that. Too often we make concepts "bigger than life" or just too complex or restrictive. Doing those to total quality severely limits what can be gained from using these concepts and tools in any safety and health program. Being a total quality program is as critical to safety or health programs as it is to American industry. Nonsurvival may be the result of not embracing total quality to American industry. To a safety or health program, adopting total quality concepts and components or not won't mean the difference in having or not having a program tomorrow. It could, however, very well be the difference in *you* being there tomorrow or not!

4

ROADBLOCKS TO TOTAL QUALITY AND THE PROCESS OF CHANGE

A few years ago, my plant received STAR status from the Occupational Safety and Health Administration (OSHA) through its Voluntary Protection Program (VPP). Finally achieving STAR status caused me to pause and reflect on the long road to that goal and to look forward to the challenges that lie ahead. I remember walking into my Plant Manager's office years earlier and suggesting that we try for the STAR. That idea met with a healthy resistance. Of course, I was more stubborn than he, and in the end, he allowed me to try. He was cautious. Who wouldn't be? He was being asked to challenge a paradigm. But this opportunity would not only be a tremendous challenge and measure of our program's strength, it would also be a benchmarking effort for our total quality program. There were just too many potential wins not to try.

Anyone who has ever gone through the VPP application, review, and site evaluation process can attest that it is no cakewalk. Many times you question why anyone would do such a foolish thing. Would achieving STAR status ever compensate you for all the work? Now having the STAR I can say, "Yes, definitely yes!" Today, however, because of being the first facility in our state, OSHA Region, *and* in the corporation to receive STAR status, we have again returned to the questioning stage. It's not the additional time demanded by the onslaught of questions about VPP; it's facing the same question I fielded earlier in our process with my Plant Manager. The question keeps coming from other facilities, "Why would anyone do such a foolish thing?"

Reflecting on the many predictable roadblocks to new ideas, such as trying for STAR status or delving into total quality, I have found that the wilder or greater the deviation from the norm the idea is, the stronger and higher the roadblocks placed in its path. Most are not bothered by the application process; nor does the idea of working in a partnership role with OSHA bother them. Commitment of time and resources is not a major obstacle. What bothers others most about the idea of trying for STAR status is the question: Why do it? Sure, it would be a feather in the cap to be a STAR facility, but why would anyone go through all that work and aggravation? In any change, the "Why" question is *always* the hardest for others to understand.

Those companies that have become quality organizations don't usually ask, "Why do it?" They ask, "What is involved?" They *know* why! Those who haven't embraced total quality, can't get past the "Why do it" question. The "Why do it" question really isn't about change or trying for STAR status at all. It's about total quality! Odd! It is also an excellent place to begin our look into the roadblocks to total quality and the process of change.

ROADBLOCKS TO TOTAL QUALITY

When introducing the idea of implementing total quality concepts, resistance is the only certainty. One of the first encountered manifestations of resistance comes in the form of roadblocks. Questions like, "Why?" or "What's in it for us?" or "What value is it?" are common. Emphatic responses are also common such as, "We can't do that!" or "It will *never* work!" or "That will gain you [or us] nothing!" or "Total quality doesn't apply to safety or health programs!" They are roadblocks placed in the path of change. Where are the roadblocks really coming from? Why do they always surface when confronted by a new idea?

Roadblocks to total quality come from four different areas: misunderstandings, paradigms, fears, and culture. **Misunderstandings** about what total quality is or applies to are very common. This is especially true if an organization has unsuccessfully tried to implement total quality—tried to implement what it thought was total quality but really wasn't and failed—if it is staffed with American business-as-usual traditionalists, if it considers total quality just a paper program, or if it has heard "war stories" about other companies and their implementations of total quality. It's a knowledge-gap, or ignorance issue. Humans have a long history of taking vehement stands with little knowledge of the issues. We respond emotionally to something simply because we are ignorant about it. A good example is nuclear power. How many people really know the issues and risks concerning nuclear power? Very few. How many are passionately against it? Many. Why? It's a knowledge issue. People take an emotional stance based on what they understand. More often than not, it is one-sided, very limited, or based on something other than the facts. It would be wonderful if the human animal would refrain from taking any position on an issue or idea until it had *all* the information and adequately evaluated it. Unfortunately, humans do not. Remember cyclamates?

One area of misunderstanding about total quality must be eliminated. Ninety-percent of the people to whom I talk about or teach total quality begin with this misunderstanding. I call this misunderstanding, which has roots in our American culture, the "Plug It In Expectation." When looking for new and improved technology or when seeking a fix for a problem, our culture has primed us to expect something we can "take off the shelf" and "plug in." By doing so, we've instantly achieved what we set out to accomplish. If our natural "Plug It In Expectation" is allowed to dictate our expectations for total quality, it becomes a significant misunderstanding. And this misunderstanding can have a large effect on our ability to see what total quality can offer. Putting this misunderstanding into words, it would sound like this: "Let's cut to the chase. Just give me the specifics, the how-to's so that I can plug total quality into my safety and health program." Total quality doesn't come that way.

Perhaps the best analogy is the difference between medicine and philosophy. Medicine is a tangible, cut it out, sew it up, fix the problem, cure the disorder concept. It fits our "Plug It In Expectation." Philosophy, however, is *nothing* like medicine. Philosophy is thought-based, highly arguable, and nontangible. Philosophy runs 180 degrees counter to our "Plug It In Expectation." That's why medical doctors are held in such high esteem (and paid very well too) and philosophers are considered to be strange.

Total quality is more like philosophy than medicine. Total quality is a way of thinking that defines how to do things and what values to have. Total quality can only be built upon this philosophical base. Merely plugging total quality tools into a traditional way of thinking will *not* work. So, if you are looking for something you can plug into your safety and health program, something that is how-to oriented, you won't find it here. That's *not* total quality, but it is a major reason many attempts at total quality have failed. Total quality is not a plug-it-in process.

The second source of roadblocks to total quality is our **paradigms**. Your paradigm may say that total quality will work and provide many advantages. Someone else's paradigm may be 180 degrees off of yours. Having a phase shift between your Total Quality Paradigm and theirs is a significant challenge. Roadblocks also can come from the other side of that paradigm, their Status Quo Paradigm. They may strongly believe that their current program is working just fine—"If it ain't broke, don't fix it." You, however, because you deal with the frustrations and failures daily, have a different Status Quo Paradigm, very different.

The third area where roadblocks to total quality come from is the most complex, **fear**. There is no such animal as a fearless person; everyone has fears. True, what each of us is afraid of and the level of that fear is very different, usually widely different. Even the most ferocious animals have fears. Why does the rattlesnake strike? Why do skunks spray? Why do elephants run away? These are natural fear reactions. When we talk about fear reaction to total quality, however, we aren't talking about large monsters. Or are we?

Don't assume that I look at fears negatively. That isn't the case. Fears are as natural to us as noses. They are useful to us: Fear is what keeps most of us from jumping out of airplanes, hang gliding off El Capitan, or soloing across the Pacific in a row boat. Fears are normal and we all have them. Second, understanding that fears coexist inside of us in balance with an attribute called caution is important. Because we live in a balance, caution is the other side of fear. Once we reach the caution area, we can deal with our fears positively.

Fears about total quality come from:

- ▸ Fear of any change
- ▸ Job security issues
- ▸ Threats to turf
- ▸ Threats to power
- ▸ Fear of multifunctionality
- ▸ Fear of measurement
- ▸ Fear of sharing knowledge

These seven personal areas of total quality fear are real. Fear of any change is the most common. Change is disruptive. It challenges the status quo, whether the status quo is good or bad. Change disturbs our environment's predictability. It's part of our nature. We love predictability. We depend on predictability. When you go home from work, what will you find as you walk into the house? When you go to tell your boss about a cost overrun, what will his or her reaction be? When you tell your son and daughter to turn off the TV and get cracking at homework, how will they respond? When you pick up your pet's food bowl to feed it, what will be its response? If your wife or husband should call you and tell you that the car just broke down, what will be your response? Predictability, we all need it in our lives. Because of this need, it is natural to be fearful of any change.

Job security fears are highly dependent upon personal confidence, history, and where someone is in the "pecking order." Job security fears, however, are one of our strongest concerns. Reading texts or talking to professionals who specialize in chemical dependency counseling, you will learn the importance of job security to people. These experts find that people who are hooked on drugs and/or alcohol will sacrifice friends, all that they own, their family and spouse before they will risk losing their job. We're a strange animal. We place the importance of a job above friends, property, family, and even our spouse. That's a pretty strong need for job security, isn't it? Fears about job security when introducing the idea of total quality, therefore, should not be taken lightly.

Threats to turf and power are as closely related as personal security or ego issues. If one is deeply invested in turf or power as a source of needed security or ego gratification, they are difficult to give up.

The fear of multifunctionality is an interesting fear. It comes from two main arenas: fear of learning and fear of competition. The unspoken fear scenario goes like this. "What will people or my boss think if I am slow or can't pick up the knowledge? I'm comfortable now. I'd rather leave it that way." Fear of competition is nothing more than the fear of losing. That, too, is a natural fear. No one wants to be a loser. No one actively looks for opportunities to lose. We select our battles with great caution, picking those where we have the best chance of winning. When you have the foreign idea to introduce total quality, it forces those who are affected by that decision to compete in arenas where they may feel that they can't win.

The fear of measurement is a personal security issue and comes from the "I don't know" bug. You know, that inner creature that says, "If they measure my performance, maybe I won't measure up." Or it might say, "I've never measured my performance before. Isn't just being busy enough?" Measurement to most people means judgment. In a total quality organization, however, it doesn't. In a total quality organization measurement means nothing more than measurement. It is hooked to expectations and performance. Now, those are two frightening words, aren't they?

The fear of sharing knowledge is a lot like the fear involved with turf and power connections. It's the classical, "If I share that information or knowledge with someone else, they won't need me anymore." Looking at it that way, to some people it may also be a job security issue.

The last source of roadblocks to ideas about total quality comes from our organizational **culture**. Culture, after all, is that nearly undefinable quality of an organization that makes it do or perceive the way it does. Culture involves the unwritten rules of an organization. For example, an organization's culture may make it more apt to take risks, or be ultra-conservative. It may make information highly visible and share it openly, or be closed and secretive. Culture makes an organization value input and ideas, or devalue anything but ideas from the top. It can drive an organization to seeking excellence, or make it content to settle for second best. A culture can convince employees that they make the world's best products and work for the world's best company. Or it can make employees feel lucky that they still have jobs. An organization's culture drives the people within it to succeed or fail and provides the rules by which either is accomplished daily.

Where does an organization's culture come from? It always begins at the top. The direction, policies, and practices of top management create an organization's culture and then continually nurture it by actions. That top-fed culture flows down through the management chain, and normally out through the staff areas until it reaches and affects each employee. Does that mean that if a company has poor, backward-looking top management it is destined to have poor organizational culture? Ninety-nine times out of a hundred, yes. It doesn't mean, however, that all areas of the organization will have the same culture. From strong leaders within the organization's structure, it is common to have radically different group cultures. For example, within a progressive, front-running organization, one department manager is a tyrant. And, because of this, the culture in his or her group would be much, much different. Consequently, it would stand out from the rest. On the other hand, within a subversive organization, a true leader can cause a positive and open culture to exist within his or her group.

Culture at any level, organizational or group, can produce large roadblocks to a total quality effort if the effort goes against the beliefs or values of that culture.

When any different idea comes along, roadblocks are common. Ideas and their resulting roadblocks, however, are only the first volley in the process of change. The idea is the beginning of questioning that results in the decision to change. The process of change involves the whole cycle from inception to completion and evaluation. If you are trying to make a change to a total quality way of business, you will have to be very familiar with this process of change. Because, by being so, when you are championing change, you can anticipate obstacles along the path and take advantage of opportunities that will improve your chances of success.

RESISTANCE AND TIME

Change, like death and taxes, is one of the certainties of life. Also like death and taxes, change is almost never welcomed with open arms. Change causes disruption and loss of predictability. Cultural changes are even more disruptive and nonpredictable. Two certainties go hand-in-hand with change. The first is that when people encounter change, it is met with resistance. Whether a spouse is leaving an abusive relationship or a company is implementing a new manufacturing process, change is always resisted. Whether it's adding grits to a family breakfast, or changing the working shift structure in your work group, change is resisted.

Years ago, I was asked to "turn around" a troubled safety program. Top management was frustrated. Compared to the entire corporation's injury rates, injury rates at this particular plant were not only high, they were out of sight! So, I was hired to change the plant's safety performance. In reality, however, I was there to change the facility's safety culture. I would, therefore, have to be a champion of change.

Because it was critical that no one in the organization think that anything was the same, I changed *everything*: written programs; accident reporting; safety committee structure and people; safety inspections; responsibility structure; safety training; the visibility of the program; and communication pathways and frequency. Change was made highly visible and unavoidable to any employee. What was the reaction? People thought I was crazy! They openly criticized the massive changes and scoffed at their chances of success. Workers and management alike began to call for return to the "old way of doing it," although they knew it didn't work. They laughed at me. They avoided me. Being very stubborn, I continued. There would indeed be a change at the plant—the safety culture or me. I got lucky—I outlasted the resistance. And yes, the safety performance at the plant markedly improved.

Did all the new programs work? Don't get hung up in the "forest for the trees" argument. To a large extent, the programs didn't matter. The change in the culture was the critical aspect to accomplishing what upper management wanted—lower injury rates.

Did openly challenging the old safety culture meet with resistance? Much more than expected. In retrospect, I see that I got hung up in the "motherhood" aspect of safety. If you are trying to make it safer for workers, shouldn't they openly embrace your efforts? Although neither management nor the workers were happy with the past safety culture, they resisted changing it. Change only offered disruption and loss of predictability. I, therefore, was a foe, not a friend. To them, it was all about the disruption brought about by change.

The second certainty about change, especially cultural change, is that it takes time. The process of change takes time. How long it takes, however, depends on five aspects, or impactors, of that change: the change trigger, the magnitude of the change, the longevity of the old culture, the size of the group or organization being changed, and the number of people working toward change.

What precipitates the change, the change trigger, obviously influences the time it takes for change to occur. For example, if a change is triggered by a do-or-die reason, and everyone in the organization is convinced that this is true, change can occur much more rapidly. If, however, the change is brought about by a good idea from upper management (better known by workers as "the change of the week"), the time needed to carry out that change will become so long that it could very well discourage the required persistence to implement the idea. This is one of the major reasons that changes to total quality take time. More often than not, the organization believes that the change to total quality is not a do-or-die situation. Thus, a change to total quality takes time and continued dedication.

Obviously, the magnitude of the change, or the perceived size of the change, can notably affect the time it takes to carry it out. An organization that completely redesigns its entire production line, makes massive changes in the organizational structure or reporting, or makes a 180 degree change in culture must expect that it will take time. On the other hand, small changes such as dropping a process step, putting a new piece of equipment into a process, adding a statistical summary to a monthly report, or changing the storage of paper clips can be implemented quickly. Changes to a total quality process are not minor changes. They are deep cultural changes that shake the very foundation of how an organization operates and turn the value structure upside down.

The time required to implement change also depends on the longevity of the old culture. Changing something in an organization that has operated in the "old way" since the ice age takes a lot of time and patience. Change in a new facility, or one that is under a constant barrage of changes, can merely be inserted into the flux.

The size of the group or organization affected by the change is also an important element that affects the time required for that change. Small groups can undergo change, even major change, much faster than large organizations. For example, look at the changes at IBM. IBM's organizational culture is defined by its size. It's "Big Blue." Because of this, organizational changes required to make IBM successful in the global economy cannot be done quickly. It's impossible.

The time required to implement change also depends on the number of people who are working toward that change. Changes championed by one person can take so long that they may very well never happen. Sure, the effect will be larger if the one champion or sponsor of change is the "top dog" in the organization. It will still be slow. Changes that are championed by many, perhaps the majority of the organization, can occur very rapidly. Look at the former Soviet Union—an excellent example of change brought about by the masses.

As the time to make a change happen is dependent upon different impactors, so are the chances of a successful completion. The success of change depends on six dynamics: the clarity of the change vision, the clarity of the reason behind the change, the ownership in both that vision and reason, the participation in the change process, the effectiveness of communication during the change process, and the prevention of covert communication.

A change vision must be clear *and* positive. Fuzzy or negative change visions are hard to sell, much less for others to visualize. It's like saying, "I want to lose weight." Okay, is a pound enough or do I want to look like I'm anorexic? The vision isn't clear, and because of that, it can easily be seen as negative. However, a statement like, "I want to weigh 155 pounds" is clear and easy to make positive. It also has a better chance of success, mainly because you know when you get there. It's for the same reason that change visions such as, "We've got to get better at quality," "We have to improve production," "We have to reduce costs," or "We have to become a total quality company," have little chance of success.

A change vision has to have a clear *and* positive reason to happen. Changing for the sake of changing doesn't cut it, nor does it have much chance of success. For example, change reasons such as, "We've decided that the present way of management is not working, so..." or "We need to get better, so..." will most often meet with failure. It's not unlike proposing marriage with a line like, "We ought to get married because I think we will be happier." With such a reason, what do you think will be the success rate of implementing that change, that is getting or staying married?

The success of change also depends greatly upon whether those affected by the change have ownership in that vision of and reason for change. It's like a husband trying to convince his wife that he should buy a Porsche to drive to and from work. If the wife has no ownership in the vision or reason (she's not going to get to drive it), the chance of getting much buy-in is limited. However, if the change vision and reason are perceived to be good for the entire organization, such as to make it more profitable and, thereby, able to stay in business, everyone will have high ownership and the chance for successful implementation of such change will also be high.

Participation in the change process is also an important factor in its success. It's a fact that people trust and support what they are involved in. Participation not only builds commitment to the change, but strong synergistic energy also takes place. Why do you think that special interest groups have been so successful in this country, groups like Mothers Against Drunk Drivers (MADD) or the multitude of environmental groups? They are successful through high levels of participation.

Communication during the change process is also critical to its success. Communication has to be frequent and effective from the very beginning of the change process to the end. Those who are participating in or affected by the change must know what is going on at all points along the path. Breaking communication results in confusion, loss of commitment, loss of direction, and feeds covert communication. These are destructive to success.

Why is preventing covert communication important to the success of change? This question is better asked another way. Have you ever heard a positive or supportive rumor?

ROLES AND ASSESSMENTS

In our discussion of change, there are usually four different roles to consider: the champion(s), the sponsor(s), the change agent(s), and those affected by the change who are called target(s) (a poor choice of words in my opinion). In a small change or change within a small organization or group, all four roles can be embodied in one person or one group. Generally, however, each role is separated and given to specific individuals or groups. As a matter of definition, a champion is the person or persons who actively push for the change and seek support for it. The champion is the visionary and the salesperson of the change. The sponsor is the one who authorizes the change. In an organization, the sponsor is usually at or near the top. The change agent carries out the change. In most organizations undergoing change, the change agents are usually the supervisor and area managers. Each role has specific responsibilities and needed skills within the process of change.

Before any change is begun, how can the chances of success be assessed? Such an assessment would be valuable information for plotting a strategy or having some idea about the chances for success. So, before a change to total quality is started, make assessments in five areas: implementation history, sponsors, change agents, resistance, and culture.

Assessments are best accomplished with questions. The answers to those questions allow one to assess the potential for success. In conducting an implementation history assessment, try these questions.

> ▸ What is the history of other implementations or changes in this organization?
> ▸ Have they been successes or failures? Has the organization at any time undergone a massive change?
> ▸ Has it had best success with many small changes moving toward a goal or with large single changes?

▸ How long has it been since the last major change was implemented?

▸ What change mechanisms were used, both successful and unsuccessful?

▸ Where did the changes begin?

▸ Who started them?

▸ What kinds of resistance did they encounter?

▸ How was support generated?

▸ What opportunities were exploited or lost?

▸ From these past changes, what anticipation and acceptance level can be expected?

Questions you can use to conduct a sponsor assessment might look like these:

▸ How can I get sponsorship from the top of the organization?

▸ How can that sponsorship be spread down through the management chain?

▸ Who else in the organization needs to be a sponsor of this change for it to be successful?

An agent of change is the person or level within the organization that implements the change. In conducting a change agent assessment, try questions like these.

▸ Where in the organization would a change agent be most effective?

▸ How can commitment best be built into these agents?

▸ What level of skills or training is needed for them to implement the change?

▸ What kind of support and communication will they need?

Anticipating resistance and positively dealing with it is key to a successful change. Therefore, a resistance assessment before starting is pretty important. Use questions like these.

▸ Where can I anticipate resistance to come from?
▸ In what order will that resistance be encountered?
▸ What form will the resistance come in?
▸ How can I keep the resistance out in the open?
▸ What avenues can be provided to communicate and quickly deal with resistance?
▸ What methods and avenues can be used to eliminate the resistance before it surfaces?

Because the organization's culture can make or break any change, a culture assessment is wise before implementing a change. Questions that might be used include these.

▸ What are the ugly elements of the organization's culture that might derail a change?
▸ What are the positive elements that might support the change?
▸ What steps can be made to turn the "ugly" cultural elements into neutral or positive ones?
▸ How will the communication pathways within the culture affect the change?
▸ What additional communication can be provided to prevent covert communication as much as possible?

THE CHANGE PROCESS

Assume at this point that we still want to move ahead with implementing a change. The assessments have also provided a lot of information and ideas about how to make the change successful. What is the best approach to implementing a change? After all, it could be the greatest idea in the world and have the best chances of success, but implemented with the wrong approach, it can only spell disaster. A successful change process is not simple. A lot of information can be found in books and courses dedicated to change management. For our purposes, let's look at a simple eight-step approach to implementing change.

Approach to Implementing Change:

1. Identify previous roadblocks to ideas of change,
2. Assess the skills, knowledge, and motivation of the key sponsors,
3. Select your key players, change agents, sponsors, champions, etc.,
4. Identify and prioritize the sources of resistance,
5. Determine the resistance from the current culture,
6. Quantify the human elements that will be needed to monitor the change,
7. Develop strategies to introduce, communicate, and implement the change, and
8. Prioritize and coordinate the use of resources and/or power.

Within any change process, there is a period of disruption that lies between the start of the change and the successful implementation of the change. That period of disruption is called the transition period. This is the period where most of the resistance is met and most of the battles are fought. It is a critical time in the process of change. During this period, remember that resistance and the amount encountered to a great extent is a function of the disruption caused by the change and the chosen change strategy. If the disruption is large, so will be the resistance. Thus, use of some standard transition management tools is important. These skills include: listening, patience, commitment to the change, active communication, information, and repeating the change message over and over.

Earlier, we briefly talked about the interrelated nature of the trigger mechanism for change and the amount of time required to complete the change. The trigger mechanism also affects how many transition management tools will be required. For example, it is generally accepted that there are four levels of change triggers: opportunity, need, discomfort, and pain. It's a sliding scale with merely having the opportunity to change as the lightest trigger and pain as the heavy hitter. This scale is inversely related to the chance of a change's success, the time required to change, and the number of transition management tools required. It is also an important consideration when contemplating a change to a total quality process.

THE CYCLE OF RESISTANCE

Resistance is an important aspect of change that is critical to know and, as much as possible, understand. Resistance is one of the certainties within the process of change. No matter what the change is, resistance will occur. Any champion of change must be able to predict the effect of resistance. The success and timeliness of the change, as well as the credibility of the champion, ride on the prediction accuracy and the proactive steps taken to counter resistance. So, let's talk about why resistance happens.

One rule concerning resistance to change, if written, would go like this: *Resistance will always run its cycle.* The speed or depth of that cycle is, of course, dependent on what we've discussed before, but resistance will *always* run its cycle. What does that mean? It means that from the start of change, resistance, which is an individual and/or group emotional response to change, will follow a predictable transformation from one stage in the cycle to the next until the cycle is finished. It can even back up in steps within the cycle, but will not be finished until the cycle continues to completion. What is the cycle? It follows in this order: denial, anger, bargaining, depression, exploration, and, finally, acceptance. Shown in graph form, plotting emotional response against time, it would look like Figure 4. Most dramatically, we see this reaction pattern in the bereavement process[5] following a death.

It is harder to see a clean run through the pattern if the change is small or the target is large, but the pattern will hold. It is also very

---

[5]E. Kubler-Ross, *On Death and Dying: What the Dying Have to Teach Doctors, Nurses, Clergy and Their Own Families* (New York: MacMillan Publishing, 1969).

common to waver between anger and depression. Any change process needs to use tools that will help surface the resistance to move it along as quickly as possible. If those tools are not used, a traumatic change can result in this vacillation between anger and depression for years. Note one critical point in the response cycle. Once the target begins to move through the depression phase, the orientation changes from looking back (past orientation), to looking forward (future orientation). This change in orientation is the key to getting through this reaction pattern and, thereby, successfully implementing change.

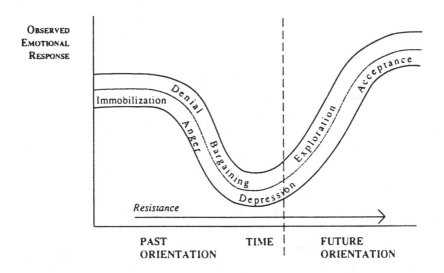

Figure 4: Emotional Response Cycle to Change

CHANGE IS ALWAYS RESISTED

Why do people resist change? This is a very important question to answer. People resist change for nine reasons. First, they resist change because they fear that they will lose control, both of what is going on in their world and also, of what directly affects them. Second, people resist because of uncertainty. This can be a huge source of resistance if a change has a foggy vision or is poorly communicated. Third, resistance occurs because of the element of surprise. This is another reason for a lot of good communication before, during, and after the change. Fourth, change resistance is a result of what is called the "Difference Effect." It is natural to resist anything that is perceived as different from what is known or historical. Fifth, people resist because they fear that they will lose face. It's an issue of misplaced loyalties or past support. Sixth, resistance comes from concerns about future competence or need. Stated, it would read like this. "If the change occurs, they won't need me any more or I won't be competent." Seventh, resistance can result from the "Ripple Effect." It is not unlike throwing a stone into a pond. There is a group synergy that occurs during change where resistance continues just because there is group resistance. The eighth reason resistance happens is the fear of more work. "If the change happens, I will have to do more and work harder with less." Again, it is a great opportunity for communication. The last reason behind resistance is past resentments. During a change is a great time to deal with all the emotions concerning past "sins" in the workplace. During change, everything boils to the top.

Change is one of the facts of life. Change brings disruption and loss of predictability to an animal that so desperately needs predictability. Total quality is change! Don't forget that! If you plan to implement total quality in your organization, group, or function, you are implementing change. Because of that, you can expect roadblocks and resistance. With knowledge, you can anticipate the roadblocks and avoid them. Further, you can predict resistance, and, through planning, minimize it. Change management is the key to bringing about a smooth, rapid, and successful transformation to total quality.

5

THE LEADERSHIP CHALLENGE

Some words absolutely stick in my craw. These words get an immediate uncontrollable reaction from deep inside, much like that in response to fingernails dragged across a chalkboard. There are just enough, however, to get my blood up once or twice a week. One of those words is "boss." Questions like, "Who's your boss?", "Are you the boss?", or statements like, "Well, you're the boss," drive me up the wall. In my mind, the word boss has a cartoon-like connotation of the near out-of-control tyrant who secretly spies on employees, makes them work late on their birthday, and never gives praise or raises. Am I the boss? With such a strong mental picture, no!

We get mental pictures from so many of the words we hear. Some create negative images and some produce positive ones. One of the words that is used to produce mental pictures of heroes has been the word "leader." Isn't it a wonderful word? It's right up there with the word "mother," or close to it. Mental pictures of a medieval king leading his armies into battle, the president of a country or huge business, or the eloquent, outspoken head of a special interest group come to mind, don't they? Whatever or whoever the image, it is almost wrapped in the flag or in divine light, isn't it? The word "leader" is indeed a neat word.

When applying one of these two words, boss or leader, to ourselves, an interesting dilemma occurs. Oddly enough, most of us relate easier to "boss" than "leader." It's driven by our mental image of both. Our modesty or lack of self-confidence just won't allow us to embrace that image of ourselves as leaders with all the accolades, honors, and respect. That's one of the reasons the term "manager" is in such wide use. Can you envision a door plaque such as, "John Howell, Leader," or "Mary Francis, Boss?" You can't envision those anymore than you could, "Fred Smith, King!" Titles get strange reactions, both from inside ourselves and from our society.

Looking at recent changes in our management concepts about business, however, the term "leadership" has been cropping up more and more. Books that used to have the word "management" in them now have the word "leadership." In the changing workplace, this wording change complements a lot of those changes. Changes that are most indicative of this are those that focus on the different ways we deal with employees. In these changing workplaces, employees no longer need to be "managed," they deserve to be "led."

CHANGE MUST BE LED

If you are thinking about changing your group's or organization's culture to total quality, that isn't a change that can be "managed." It *must* be "led." So, in our total quality challenge, the word "leadership" is a very appropriate term. One that we should become much more familiar with. Why? Because, there are no such things as born leaders. Nope! Just like there are no such things as born corporate presidents, military generals, or leaders of change to total quality. The all-too-common phrase, "Born Leader" is a misnomer. It just doesn't happen! Leaders, like chefs, firefighters, and lathe operators, *learn* how to be what they are. So, if leadership is learned, it is a good subject to discuss here, and *learn*!

"You've got a great program," a fellow safety practitioner said after coming and taking a look at our safety program. "I wish that we could do this at my plant," he continued. "You can," I encouraged. "It doesn't have to start with a vision from God. Anyone, including you, can be a champion of change and begin the transformation, if nowhere else, in your own group. I've seen it happen many times."

"Oh, I can't do that! How would I go about it? I just don't think I have what it would take to lead a change like that!" he concluded. This conversation speaks eloquently of why the subject of leadership is so important to our discussion of total quality. Because, whether the focus is a program, a group, a department, or an organization, total quality *is* change. And, change must be led if it is to be successful.

THE ELEMENTS OF LEADERSHIP

What are the traits of a leader? First, leaders don't *look* different from the rest of us. They just *do* things differently. Books on leadership provide a lot of different perceptions of leadership. The description of leadership I like best and will use here is described by Kouzes and Posner[6]. From their work, they describe five elements of leadership. The five elements that leaders display will be useful to our discussion of leadership. They include: challenging, inspiring, enabling, modeling, and encouraging.

The element of leadership, **challenging**, if expanded into a sentence form, might read like this: "A leader challenges the status quo and the accepted ways of thinking and doing things in an effort to make improvements." A leader is constantly "stirring the pot." This element runs 180 degrees counter to traditional management concepts. Management maintains the status quo. Leadership questions and improves it. Challenging is exactly what you are doing right now. You are taking an in-depth look at the concepts and tools of total quality and seeing how they might serve your program or organization. That's challenging the status quo. What my safety practitioner colleague mentioned earlier in this chapter didn't recognize was that by coming and seeing my program, he was already exploring one element of leadership. He was challenging the status quo. Too often, like him, we become myopic about our own abilities and actions.

---

[6]J. Kouses and B.Z. Posner, *Leadership Practices Inventory*, (San Diego, Calif.: Pfeiffer & Co., 1988).

The second element, **inspiring**, is a group action. After all, inspiring yourself is extremely difficult. You're either turned on to something or turned off to it. Inspiring, therefore, is the ability to turn others ON. When we think of the ability to inspire, however, too often we move it into the near metaphysical realm. The person has such presence and skills of persuasion, that all around can do nothing but follow, glassy-eyed. Thank goodness that few have that kind of presence. Otherwise all of us would have whiplash from changing directions all the time. This example, however, is *not* what I mean by inspiring. If it were, our society would have few leaders. The trick in inspiring others is, first, to know what communication pathway to use, second, to know what their communication needs are or what language to use, and third, to have enough communication and presentation skills to effectively sell your product. It isn't metaphysical or something you are born with. It's a learned ability.

Enabling, the third element of leadership, is also a group activity. More specifically, enabling talks about how we work in and with teams or groups that work together to perform a specific function. Enabling has multiple facets. It fosters individual ownership, not personal accolades and turf building. It reinforces the positive actions of others. It helps. It doesn't get in the way. It is supportive and understanding. It is not judgmental, discouraging, or condescending. Enabling builds the leaders of tomorrow.

Enabling is like the father or mother who helps his young son or daughter learn how to ride a bicycle. Together they make the bicycle ready and talk about what part does what to make the bicycle go. They talk about how to ride it. All questions are answered quietly and in great detail. Perching the youngster on the seat for the first time, the parent slowly begins to move the bicycle while holding the handlebars tightly *and* the child gently. While using soothing and reassuring speech, the parent smoothly propels the bicycle. Gradually the bicycle and child are accelerated to a speed that is comfortable to the child. Still holding on, the parent weakens the grip and lets the child gradually feel what balance is all about. Slowly, and at the child's pace, the child takes control of the bicycle. All through this, the parent gives encouraging words with a watchful eye and continues to soothe the child's concerns. When the child solos for the first time, the parent celebrates the accomplishment with the child. That's enabling!

The fourth element of leadership, **modeling**, could best be phrased as "walking the talk." Leaders don't talk out of both sides of their mouth. They stand upon the same business values and work ethics that they exhort others to stand upon. And, more important, their actions match their words. Of all the elements of leadership, this one gives the most heartburn. Why? It does so for three reasons. First, modeling requires that you first have values and good ethics, and an acceptable level of work productivity and quality. This in itself is ambiguous. What level of value is appropriate for a leader? Surely, it must be higher than that of someone who is not aspiring to be a leader. This is a perception trap that you should steer clear of. It is not level-dependent. The second reason is that, usually, these are not learned skills. Rather, they are acquired though your history, past successes and failures, the organization's expectations and culture, etc. And third, through modeling, a leader can lead in any direction. If the leader has low values and a poor work ethic, then that's the direction he or she will lead. If, however, the leader is an idealistic perfectionist, the impact may be exactly the opposite.

The last element of leadership could, in some ways, be considered part of enabling. Because of its importance in almost everything a leader does, however, **encouraging** is set alone as an element of leadership. Like enabling, encouraging has multiple perspectives and can best be understood as "encouraging the heart." A father or mother figure, a coach, a sports fan, or a team member all encourage, but all encourage differently. Each encourages from a different perspective and role and provides for different needs, different levels of respect, and different levels of expectations. Encouraging is no more than being a cheerleader for individual and team effort. It is never negative. It is never abusive. It ignores failure, seeking only what is learned from it. It rewards effort, not just success. Fortunately, encouraging is usually not a learned skill. We already know how to encourage others. It's natural to us. Encouraging is usually a matter of awareness. If we know how important it is, and we notice our opportunities to encourage others, we can consciously change our behavior. Encouraging is the *easiest* element of leadership to master.

WHERE ARE THE LEADERS?

A colleague of mine said once, "The most effective leaders are most often not the ones at the top." The truth of that statement still sends chills up and down my spine. In any organization or effort, if success solely rested on the leadership of the person at the top, most, if not all organizations and efforts would fail. Success depends on leadership being strong at the middle and lower levels of the effort or organization.

Whenever we talk about change and leadership, we need to talk about the darker side of the safety and health profession. Historically, as change has occurred, safety and health professionals have tended to resist change and stay on the outside rather than being leaders in that change. Perhaps it's because we set ourselves apart from working types and other staff functions. Maybe it's just because we are more conservative than other disciplines.Whatever the reason, when change occurs, we take a comfortable back seat and watch it succeed or fail. We have held too long to the status quo, the way it has always been done. Total quality offers one of those rare opportunities for safety and health to lead the charge. Taking the initiative to change, however, requires us to assume a leadership role and to embrace change instead of sitting by and watching. This difficult challenge is one that we must accept as necessary for the future success of our mission *and* us.

Take heart! Leadership is a skill that *anyone* can gain by honing his or her skills and awareness. It isn't a "born with or without" part of the body or brain. It isn't mystical or magical. It *is* dynamic and exciting. Anyone considering a change to or into a total quality process must lead that change.

Are you a leader? My safety professional friend had concluded that he was not. Too often we make judgments, especially about ourselves, without adequate information or knowledge. Now you know. See? Leadership is a piece of cake.

6

LINE VS. STAFF: SHIFTING THE FOCUS

Can you imagine a world in which all parents refused to perform a particular duty in raising their children? Let's say that all parents refused to help or teach their children; or they refused to correct them when they made poor choices or did wrong. Let's say that they stopped serving their children meals or providing a place for them to sleep. How about if all parents refused to provide clothes for their children or to send them to school? Can you imagine the havoc caused by any one of these mass-refusal actions? Society is just too dependent upon the majority of parents performing their child-raising jobs well. All aspects of it. Too much of a breakdown causes large ripples in, as well as great costs to, society. Looking at the large costs that society pays today for the shortcomings of a few parents, can you imagine the impact if *all* parents refused? Of course you can't! That society could not survive; it would unravel from its own inefficiencies. Because of the costs and disruption, it would cease to exist.

MANAGEMENT RESPONSIBILITY

Reviewing this analogy gives me cause to pause and reflect on what, for many years, has become common in America's workplaces—line management not required to shoulder its full responsibilities! "That's a pretty bold thing to say," you might think. Let's back that up with some discussion of what management is theoretically responsible for accomplishing. If you look at the total spectrum of management responsi-bility, it's divided into five basic areas: production, quality, costs, morale, and safety. In today's regulatory world, often you will also see a sixth area of responsibility, environment. Historically, and even extending into today's workplace, one area, above all, has been allowed to "fall off the table." Thus, it's treated as the least important of the five. In most cases, it is even delegated to others. That area is safety. If production falters, the supervisor or manager is "called on the carpet" for his or her failure. If quality is poor, the manager corrects it *immediately*. If costs are too high, upper management will "hold their feet to the fire." If employee morale is low in a work group, the supervisor is consulted with immediately, helped, and encouraged to turn it around as quickly as possible before it affects the other areas—production and quality. If safety goes to pot, what happens? Management blames the safety department! What? Hold the supervisor or manager responsible for safety in his or her own area? Production, quality, costs, and morale, yes, but in most American workplaces, safety and health responsibility is misplaced. More often than not, this failure to assign safety responsibility properly is painfully real.

Why has this happened? If you go far enough back in time, back to the industrial revolution in America, it used to be the employee's responsibility to "work safely." Hazards in the workplace were accepted by both management and workers as "conditions of employment." That was an extremely sad time in our nation's industrial history. In some areas of our business environment today, that belief, hopefully to a lesser extent, still holds true. For example, even with the hazards and degree of change constantly encountered in the construction industry, this belief in worker responsibility for safety still prevails in many construction firms. It's one of the reasons the injury rate in construction is so high.

However, we need to shoulder some of the blame ourselves. Another reason management responsibility for safety is not enforced is, to a large extent, our own fault as safety and health practitioners. In building our turf and desiring job security, we have willingly expanded our scope of responsibility and not allowed it to be placed where it belongs—on line management. I call it the Safety Responsibility Paradigm. Think about it. How many safety programs in industry do all the accident investigations, do all the safety inspections, do all the safety training, and track the corrective actions? It's common isn't it? Case closed!

LINE VS. STAFF

Looking at the full spectrum of line management responsibility and how we, in our conversion to total quality, can shift the focus, let's begin very basically. Let's begin by looking at the traditional line and staff organizational structure. Whether talking traditional management or total quality, this structure hasn't changed much. Historically, a line organization is pictured as a pyramid. The Chief Executive Officer, CEO, sits at the top and the management chain expands as it moves to the bottom of the pyramid where the whole organization rests on the workers, figuratively *and* literally. This translates to a direct reporting structure from the bottom to the top. The worker reports to his or her supervisor or foreman. The supervisor or foreman reports to the general foreman or area manager. The area manager reports to the department manager or superintendent. The department manager or superintendent reports to the production or plant manager. The production or plant manager reports to the general manager. The general manager reports to the vice president. The vice president reports to the president, and the president reports to the CEO or the Chairman of the Board. This is a pretty standard line structure. The position names may be different and the levels of command may be greater or fewer depending on the organization's size, but it follows this pattern. Line structures, routinely, have two legs in this pyramid: production or operations and maintenance.

Staff organizational structures, however, are flatter and much more complex in reporting lines of authority; they also have much more variability in structures and staff roles. Staff function and reporting are tailored to meet the needs of the organization or level of the organization. Most common staff groups, however, include safety and health, environment, quality assurance (or control—a poor choice of words), engineering, human resources (a confusing choice of words) or employee relations (another confusing choice of words), purchasing, marketing, and accounting (or controller—a very appropriate choice of word). Reporting structures in staff functions also vary. For example, in some organizations safety and health report up through human resources. Occasionally, however, safety and health reports directly to the facility or plant manager. In a few organizations, safety and health report up through environmental, quality or engineering. So, reporting structures for staff functions are a "mixed bag." Most often, they are custom fit for the organization and the organization's needs and priorities.

TRADITIONAL SAFETY AND HEALTH RESPONSIBILITIES

Whatever the line and staff organizational structure, traditional "old school" responsibilities are placed on the safety and health department or function. Those responsibilities would include making and interpreting safety "rules," assessing or policing compliance with the "rules," ordering and/or issuing safety equipment, performing or coordinating hazard evaluations or analyses, providing safety and health training, keeping injury statistics, and maintaining safety and health records and files. Multiple functions such as auditing and investigations would be included in one or more of these responsibilities.

Several notable pitfalls in these traditional staff-type safety responsibilities bear discussing. First, and near the top of the list, is the safety and health person or group having responsibility without authority. This pitfall is the greatest source of frustration and anxiety to safety and health professionals. A great example of this falls under the responsibility to assess compliance with safety rules. As deficiencies are found by safety and health personnel within a formal or informal safety and health audit, because no direct authority exists over that function, a "paper-trail" begins. And, like the pink Eveready® bunny with the bass drum, it keeps going, and going, and going. Safety and health people frequently use the phrase, "Document your findings," which is better phrased, "Cover your ass!" This phrase, no matter which version you prefer, speaks directly to this pitfall—responsibility without authority. We in safety and health should *wake up*! If nothing else causes us to challenge our Safety Responsibility Paradigm than this continually used phrase, it should be enough to rocket home the point that the responsibility is placed on the *wrong* position!

The second pitfall found in traditional safety responsibilities is a significant factor that makes the job impossible—I call it the "Omnipresence Confounder." Because safety responsibility falls on a single person or small group of people, they face an inescapable fact concerning space and time. They cannot be everywhere all the time! Knowing that a responsibility is impossible before it starts only brings frustration and resignation. It should, like the first pitfall, scream of the painful reality that the responsibility is misplaced.

Third, these traditional safety responsibilities produce significant negative perceptions at all levels of the organization. In these organizations, upper management has a safety and health person or group because it documents their commitment to worker safety and health. You know, "Sure we're committed to worker safety and health. See! We have a person (group) that assures workers are safe and healthy." So, the prime reason for the existence of the safety and health function is because they've *got* to have them, not *choose* to have it. Too often the function's existence is only driven by corporate directive or policy. From middle management's perspective, it isn't much better. Because middle managers haven't accepted safety as their responsibility, they look at the safety person or group as having no value unless it serves their purpose. Safety becomes an employee whip! When safety does a safety audit, they are only there to create work and additional costs to the department!

You would think that safety and health under this traditional approach would be heroes to, or at least appreciated by the workers. Wrong! In this classic approach to aligning safety responsibilities, workers view safety as the person or group who makes those complex, incompre-hensible, and unnecessary safety "rules." That isn't all, is it? Workers also perceive that safety is only there to hassle them. After all, if safety personnel would just stay in their offices and off the shop floor, then workers could really meet their production goals. Too often, unfortunately, because of frustrations or fear of conflict, safety personnel do just that. They stay in their offices!

The fourth pitfall in this traditional view of safety responsibilities results from the first three. The ugly thing to the safety and health practitioner who exists in this traditional responsibility world is the perception of little importance. Because of that perception, what is commonly the first staff function that is "downsized" or "adjusted" in a lay-off or reduction in force (RIF)? Often safety and health.

SAFETY AND HEALTH IN A
TOTAL QUALITY ENVIRONMENT

We've looked at the "ugly side" and the problems involved with the traditional approach to safety responsibility alignment. We've also looked briefly at some of the possible triggers or "wake-up calls" that should bring about a change to total quality. How then is life different in a total quality safety program? How are the responsibilities and perceptions different?

A total quality safety program totally reorients the safety responsibilities. It could be called a total phase-shift from the traditional approach. The safety person or group is no longer the safety and health "doer." Under total quality, safety is a team concept. The safety person or group becomes a helper and facilitator. If put into a sentence, it would read like this: *"Safety is a line responsibility."* In a total quality program, this is not an option. It is not a choice item. *There is no other way that safety responsibility can be aligned.* Trying to implement total quality concepts in a workplace that thinks safety responsibilities in a classical, "Old World" way won't work! Safety, like production, quality, costs, and morale must be a responsibility of the line structure. And like the other four areas of management responsibility, the line structure must be held accountable for any problems or deficiencies in the safety area.

You've probably heard, "Safety is Number 1," at least a thousand times. The question is does anyone *really* believe it? Or is the purpose of this phrase merely a smoke screen or a wishing-and-hoping statement?

Numerous industrial sites that openly brag that safety is first, have banners and signs everywhere saying so, but choices between production and safety are answered the same way, "Do the job as safely as you can, but get the product out." So, which is first, really?

Total quality programs perceive this situation a little differently. There isn't a need for competition between responsibilities. Safety has to be in a *balance* with production, quality, costs, and morale. Elevating any one element above the rest throws the whole equation out of balance. It forces one to choose between the others based on circumstances, situational issues, or perceptions. It is a no-win situation. Balance maintains perspective. It allows for difficult questions without the potential for compromise. It allows for open discussion of perceptions and issues. In total quality, safety is *not* number one. Neither is production, or quality, or costs, or morale. It's a simple relationship based on mutual dependency. For example, without production, there is no need for safety. The business just won't be around that long. Without quality, there is also no need for safety. The company may be in business a little longer, but in today's dog-eat-dog world, not for long. Without cost control, there is no need for safety. You can't afford to stay in business (unless you're the federal government). Without good employee morale, no company can have good safety, make production levels, or have excellent quality. All must be kept in *balance* as a *line* responsibility.

In a total quality safety program, the safety department, whether it be one person or a hundred, is a service-type organization. What does this mean? Safety exists to serve its customers. Who are safety's customers? Everyone in the organization is its customer. This allows a conversion in thought and perceptions to occur about the safety person or group. No longer are they rule makers and enforcers. In a total quality program they provide services to the line organization so *they* can better perform *their* safety responsibilities.

Because in a total quality program safety is seen as line responsibility, greater emphasis is placed on employee participation and empowerment. The employees' role is critical! It extends into a lot of the responsibilities that are on safety (or captured by them) in the classical scheme of things. In total quality, employees are the primary persons in the area of identifying *and* correcting hazards. Safety inspections and audits are no longer the prime method of finding safety problems. Day-to-day, hour-to-hour observation by employees is! The employees *must* have this responsibility. Employees also must have "line stop" power. This extends into production, quality, *and* safety concerns. If something isn't right, the employee must have the authority *and* freedom to exercise a "line stop." Using two simple mechanisms such as these, the employees become the agents of change within the workplace. If problems exist, the employees take ownership in correcting it. Grassroots ownership in change and correcting problems is *very* powerful.

One of the prime responsibilities of the safety person or group in a total quality program is to organize and standardize the total safety effort. This responsibility extends into coordinating or mentoring special projects, serving on review or problem solving teams, sharing information and ideas between areas and departments, and implementing facility-wide safety programs. Within this responsibility, the safety person or group also has the task of measuring the safety effort. Measurement is critical to the success of *any* program. Whether a safety program is classical or total quality, injury rate information is tracked. However, in a total quality program, different measures become more important than injury rates. One is safety participation.

Communication becomes a very important responsibility of the safety person or group in a total quality program. This includes not only active communication to various levels or areas of the organization but also to enlarge communication channels. Communication in a total quality program is very important. It must flow up the chain, down the chain, and between areas, departments, and functions. The service role of safety in helping communication and safety's champion role of total quality can be important catalysts to any total quality organization.

In a total quality organization, the perceptions at all levels are also vastly different from those in a classical safety and health program. Because of the alignment of safety and health responsibilities within the line structure, upper management perceives the safety person or group as adding value to the organization through service and information. The importance of the safety group is higher because of this. In the minds of middle management, the safety person or group is a provider of valued services, *as needed*. Safety's change in role from police officer to helper and the acceptance of safety as a line responsibility are the major reasons for this change in perspective. The workers also have a different perspective about the safety person or group. In a total quality organization, the safety person or group becomes the champion of the workers' safety. The employees are the change agents while the safety group is the champion of *the employees'* actions.

SAFETY AS A LINE RESPONSIBILITY

"Safety is a line responsibility," is more than just a catchy phrase. We've heard it many times. In organizations that use the classical, "Old World" concept of safety responsibility, you hear it, but it is really just "lip service." The placement of the safety responsibility away from the line structure alone makes the phrase a lie. In a total quality safety program, this phase takes on life and an expanded meaning. It translates from the top of the organization, down through the management chain, and into the empowered workers. Empowerment energizes it.

Many rewards result from shifting these safety responsibilities into the line organization, and there are many rewards. These rewards come to the safety and health practitioners, the total safety and health effort, management, and workers. A synergy occurs with shared importance of safety, production, quality, costs and morale, and all thrive on each other. Each supports the other. In a total quality system, you simply cannot have bad safety and health with good production and quality. If you have poor safety and health, you also will have poor production and quality. Because of the interrelationship between all, conflicts between camps disappear. Decisions become easier and not one-sided. Communication improves because the fears concerning frank discussions are greatly reduced.

How does your organization *really* assign the responsibility for worker safety and health? Want an easy way to find the answer? Think back. Who was called on the carpet the last time your organization experienced a serious accident? Who was held accountable for it? Was it the area manager, the involved worker, no one, or *you*?

7

EMPLOYEE PARTICIPATION, EMPOWERMENT, AND OWNERSHIP

Why are we always surprised when we read about some employee suggestion or idea that revolutionized a manufacturing process, saved a plant from closure, or turned a company from being a loser to a big winner? "How about that! Boy, that's really unique!" The shocked reactions we give always amaze me. Did we just fall off the turnip truck? Were we all born yesterday? Did we honestly believe that most of the good ideas in business come from the top? Get real! If you took a poll on the origins of revolutionary ideas in business, you would find that the great majority, perhaps 99 percent of them come from a creative, empowered employee, *not* from the top!

Taking a very unscientific look at the origin of great ideas in different businesses, one day a group of safety and health consultants made this analysis of how the data would look if a study were done. They did so dealing strictly from the combined experiences (more than 200 years) of the group. The results of the experience-based study indicated that if businesses were divided into two groups, those that are successful and those that are not, the origin of great ideas as well as the number of those ideas would vary greatly. In struggling or unsuccessful companies, great ideas that "saw the light of day" would be few and as much as 80 percent of them would come from halfway up the ladder or higher. The reason the business could not succeed would be because it couldn't generate ideas at a rate equal to the competition. Big surprise, huh?

However, in companies that were obviously successful and leading their fields, there would always be too many ideas. Ways of prioritizing those ideas would be necessary. The origin of those great ideas would come mostly, estimated as high as 90 percent, from the grass roots of the business, not from the top, not from the middle, not even from the engineering departments. Surprising? No, not really. Anyone who keeps his or her eyes open can see the same dynamics in his or her own organization or experiences. Then why are we always so shocked to hear about a revolutionary idea coming from a worker?

Too many times, we as safety and health practitioners and line managers forget this fact of business. We get lost in the concept that workers are there to make a product, fix equipment, inspect for defects, or ship products. We focus on them as "worker bees" and forget their creativity. And when we do so, we not only use about 20 percent of the workers' abilities, we also make our jobs harder. In doing so, we cut all of our throats! It's one of the painful aspects of "Management Myopia."

A COST OR A RESOURCE?

Two very prominent CEOs were talking about business, as they saw it, in the twenty-first century. The way they approached one subject revealed a lot about their priorities in business and whether or not they had a good chance of being in business in the new millenium. The subject was employees. One CEO said, "Employees are our most costly resource." The other said of the same subject, "Employees are our most valued resource." Do you see the difference? Sure, it's a "glass half-empty" versus "glass half-full" argument, but the wording makes *all* the difference. The word "cost" is a negative concept. When the first CEO described employees as a cost, he might as well as talking about his facilities' monthly electric bill. For example, when your spouse comes home with a new dress or a golf club, you ask, "How much did it *cost*?" What you are really asking is, "How much of a hit did the checkbook or charge card balance take?" The other aspect of costs is that when we talk about them, we are always looking for ways to reduce them. That's why the words "on sale" were invented. "It's the *cost* of doing business." "How much will that budget item *cost* next year?" "Wait till it goes on sale then it will *cost* less." "How can we reduce the number of employees to reduce our *costs!*" The word "cost" is a negative term.

The other CEO, however, used the word "value." His choice of words stressed the positive. He didn't focus on the unavoidable costs of having employees. He focused on the value that employees returned to the business. The word "value" is a positive term. The person with this positive attitude concerning employees invites ideas, listens to them, and uses them to improve the process. The use of the word "value" versus "cost" tells a lot about how important employees are *and* how they are used.

From a historical perspective, American business is fueled from the top of the organization. It's a top-down approach to conducting business. The top of the organization drives policy, direction, and energy down through the layers until the responsibility rests on the shoulders of the workers. A total quality business operates to a great extent in opposition to this historical model. In a total quality business, the energy and ideas come *up* through the organization, guided by a common vision and mission from the top. Management enables this to occur and becomes responsible for acting upon or enabling the valuable ideas of the workers. It's a top-down and bottom-up, two-way approach to conducting business. Direction, in the form of a shared vision and mission, passes down through the organization while ideas and energy flow up. Obviously, total quality requires a high level of participation by employees. And through providing opportunities, an environment for employee empowerment, and sharing ownership in the organization's efforts and successes, total quality gets energy and makes it successful.

EMPOWERMENT

What is empowerment? No, it's not hooking someone up to a 12-volt battery. In many ways, though, the figurative results are the same. In horse racing, another way of saying empowerment is, "Giving the horse its head." It's the act of letting the horse run as fast as it can go, not reining it in or steering it one way or the other, just letting the horse go as fast and in whatever direction it chooses within the wide path of the race course. What does the jockey do? Gets a firm grip and holds on for the wild ride. With a fully "empowered" horse, many times the jockey may want to interfere and influence the direction or, in fear, slow the animal's speed. A good jockey, however, doesn't. He or she knows the importance of not doing so. The jockey gently steers and hangs on.

This is a good analogy to describe empowerment in the workplace. In an empowering environment, employees are "given their heads." It releases their creativity and knowledge to run as fast as they can, in whatever direction they choose within the described path. The described path is set by the organization's shared vision and mission. Management's job is to gently guide and hold on for a wild ride. As with the jockey, management may want to interfere with the process by affecting direction or speed of action. Good total quality managers, however, refrain. Too well they know the value of the unrestrained environment that they have helped to create. That's worker empowerment!

OWNERSHIP

What's ownership, then? Is this another way of saying profit sharing or holding stock in the company? Too many top executives think so! No, ownership is *not* a financial concept. Nor is it actual physical ownership such as owning a car, a snowmobile, or a bicycle. It is a deeply personal concept. Let's say that you are planning a vacation at the beach. Looking in the mirror, you see some extra bulges that don't look the way you want them to when you are on vacation, in public view. Your vision is a slimmer, trimmer self. So, you commit yourself to taking off some of those extra pounds and shaping up. And in the time before your vacation, you work hard at shaping up. You cut back on desserts. You reduce your helping sizes at meals. You pass on the potato chips. You drink water or low calorie drinks instead of beer or regular soft drinks. You exercise regularly. As the time of the vacation grows closer, you watch the pounds melt away and your appearance in the mirror more closely resembles what you want. At vacation time, you fit into that dress or short size you wanted to make, and you look the way you had envisioned yourself when you started to get your body ready for your vacation. Having a clear vision at the beginning and having a personal interest in the outcome, you have ownership in that outcome. Ownership in a thinner and trimmer you kept you dedicated to the task and rewarded you by your success. Without a vision of what you wanted or having any ownership in the desired outcome, you would have had no chance of achieving the results. The issue of ownership is no different in the workplace. It's not a financial concept, but a deeply personal one.

PARTICIPATION: WHERE DO WE START?

Let's talk about the meat of this chapter. How do you build empowerment and ownership through participation? After all, knowing the virtues of empowerment and ownership, the "how to" is important, right? I'm going to limit the scope of the rest of this chapter to applications in safety and health only. In discussing the "how to's" then, we need to break our discussion into three areas: training, team participation, and individual participation.

"He's an idiot! And, it makes no sense at all to ask an opinion from an idiot!" A supervisor responded to my question concerning why a particular employee was not asked to participate in a problem-solving team. "He may very well be, but why do *you* think he's an idiot?" I asked the supervisor. "First, the ideas that he usually comes up with are so weird, not even practical. Second, he doesn't know anything about that process. And, I know what you're going to say. I just don't have the time to train him on the process. Besides, he's a slow learner, you have to explain everything to him over and over again." I knew this employee from experiences working with him. True, he could ask some of the damnedest questions and he was like a dog with a slipper in his mouth. Once he got his teeth into it, you could shake him to death before he would let go. It was also true that he was creative and more than amply stubborn. Was he an idiot and a slow learner? That wasn't how I knew him to be at all.

"John," I began my rebuttal with the supervisor, "he's the most creative person you have. That's probably the reason his ideas seem so 'off the wall' to you. Him, a slow learner? I don't think so. He probably just wants to learn a new process very well and feel comfortable with it before he lets go of the learning opportunity." "I still don't have the time to train him," the supervisor countered. "How much time have you lived with this process problem you're trying to solve, and how much time are you willing to chase ideas from your problem-solving team that may not be as good as the ones this employee can come up with?" I responded. "Hmm...you've got a point there," he said.

This story shows about how, all too often, we in management and in safety and health deal with employee participation. It is too easy to throw roadblocks in the path. Some roadblocks we create in our own minds and some we allow to happen by our own inaction. Each of us will always fight the "who do I want involved" and "how much time and effort am I willing to invest" monsters in our own minds. However, the largest roadblock that we place in the path of progress is due to our own inaction. It's what we don't do that hurts the prospects for participation the most. And the area of inaction where we fail the most is in training. We simply don't teach our people how we want them to participate and give them the skills so they can.

It's not surprising that we fail to do that. We are a product of our own development. After all, how much training on developing or managing a safety and health program is normally given to prepare a new safety and health program manager or administrator for his or her first challenge? Usually, it's a course or two in college, however long ago that might have been. Maybe they got a little experience standing in for the boss. It's a travesty that virtually *no* training is given to new heads of safety and health programs. I guess it is expected to sink through by divine osmosis or something.

It's no wonder that we make the same mistake when preparing employees for participation. We tell them that we want them to participate or assign them to a team, tell them what we want them to accomplish, and then we're surprised that they won't participate, can't work together, take too much time, are ineffective, or come up with crazy ideas. That's when the "boo-birds" come out—"They're all stupid." "Teams just don't work." "Those idiots are incapable of any constructive challenges." We, in fact, cast a fatal flaw into the equation before it even started. We didn't adequately train them.

If training is that important, what kind of training do they need? Start by asking yourself two simple questions. "What do I want them to participate in?" And, "What skills will they need to do that?" The following table of participation "desires" and their associated training "needs" is provided for your reference. It is by no means inclusive. It illustrates the question-answer process that must be done.

Desired Participation	*Area of Training Needed*
Identify Hazards	Types of Hazards
	Causes of Accidents
	Observation Skills
	Communication Skills
Problem-Solving Team	Problem-Solving Technique(s)
	Communication Skills
	Working-in-Teams Skills
	Process Familiarization
Change Planning Team	Change Dynamics Skills
	Planning Skills
	Communication Skills
	Working-in-Teams Skills
	Group Dynamics Skills

Training is critical to participation. You can't assume that any employee, even in management, possesses the skills necessary for effective participation.

PARTICIPATION IN TEAMS

Team participation brings with it some unusual challenges. These challenges include individual traits and strengths of team members, mix of individuals assigned to the team, official and unofficial teams, team mission clarity, team restriction clarity, team resources, time provision for and coordination between team members, timetables of expected team milestones and completion, team mentoring and sponsoring, and team communication pathways. Team participation is not as easy as "falling off a log." Because the purpose of this book is not to teach team dynamics or how to build teams, we won't discuss these team challenges. But, before you use teams to any extent or in complex challenges, you first must become familiar with these challenges and how to use teams effectively. A lot of team-building texts are available, and I've included some in the references and bibliography.

In total quality safety and health programs, teams are very effective in two areas: problem-solving and oversight or coordinating efforts. Problem-solving is either directed at a process or at a particular problem. Process-directed teams work on making a process more efficient, e.g., reduce cycle time, reduce product travel, reduce process steps, simplify work cells, maximize the use of people, perform a process hazard analysis, etc. Process-directed teams usually work within the "givens" of a particular system, such as Just-in-Time. Working with the elements of JIT and focusing on the seven deadly wastes, a process is improved and hazards eliminated. For our intended purpose, note that process improvement is *not* only in the manufacturing sector. Process-improvement teams also can focus on service functions such as purchasing, providing engineering services, approving chemical products under the Hazard Communication requirements, responding to chemical spills or fire emergencies, doing accident investigations, taking industrial hygiene samples, readying and shipping hazardous wastes, performing medical examinations, etc.

The other area where teams are used in total quality safety and health programs is in solving identified problems. These teams usually use a particular problem-solving technique such as CEDAC[6], Analytic Trouble Shooting[7], etc. This speaks very clearly about the need to train all team members on the particular problem-solving tool used. In a total quality program, it is common to use problem-solving teams for focusing on manufacturing problems. Additionally, participation of safety and health personnel as active team members is critical to helping guide the group's efforts and avoid safety and health concerns in the solution. Problem-solving teams are not used to focus on specific problems or inefficiencies in safety and health programs. Problem-solving teams are not just for manufacturing or quality purposes. Participation of safety and health professionals in team problem-solving efforts and the focus of teams on safety and health problems offer great rewards in a total quality organization.

INDIVIDUAL PARTICIPATION

Individual participation in total quality safety and health programs also can provide distinct advantages over traditional safety and health approaches. Under traditional safety and health beliefs, if the job of safety and health wasn't performed by the safety and health professional or group, it just didn't get done. In a total quality program, all employees at all levels are responsible for safety and health. It is through this individual commitment and participation that real success in worker safety and health is made. Comparing the results to those under a traditional program approach, a total quality safety and health program will produce a quantum difference. The reason is simple. In a traditional system, only 1

---

[6] CEDAC is the registered service mark of Productivity, Inc.

[7] ATS Kepner Tregoe is a registered trademark of Kepner-Tregoe, Inc.

percent or less of an organization works on identifying and correcting hazards and behavior. In a total quality program, 100 percent of the organization does.

What kind of safety and health responsibilities can be spread across the organization? It would probably be easier to say what can't be? For our purpose here, let's briefly discuss four of the most successful areas.

The first area is in making safety observations and providing or arranging for corrective actions. This is not only a participation tool but also increases the safety awareness across the organization and builds observation skills. It "tunes the senses." Employee empowerment in the area of observing and correcting safety and health concerns is a powerful tool for building employee ownership in corrective actions. Many programs have been used to build this participation in observation and correction. One of the most notable examples of this type of program is the STOP™ program marketed by DuPont. STOP™, the Safety Training Observation Program, has been used successfully in industry for many years. Other corporations have their own programs but few have been as successful at building observation and corrective action skills than the DuPont STOP™ program.

The second area where safety and health responsibilities have been successfully given to workers is in audits, inspections, and investigations. Employee participation in these activities also builds safety and health skills and knowledge. Normally, these are done in a team structure. However, individual participation in audits and inspections is also extremely valuable to a total quality safety and health program.

The third area is in safety research and vendor communications. These are both outside focused areas of participation. Doing safety and health research, like finding a particular piece of personal protective equipment or a new design for an automatic conveyor or ergonomically designed hand tool, builds knowledge of the processes, hazards, and mechanisms for corrective actions. Vendor communication responsibilities are successful because they build communication skills and focus the talk between the vendor and the employee who is most familiar with the problem. Therefore, the problem-solving solutions are usually more effective, less restrictive of the employee's job, and more accepted by fellow workers.

The last area where safety and health responsibilities have been successfully given to workers is in safety presentations and displays. Too often in traditional safety and health programs, the safety department or area management puts up a "Work Safe" poster or types up a "canned" safety talk and expects the talk to build safety awareness, correct behavior, or improve participation. Ninety-nine percent of the time they don't work. One hundred percent of the time they're perceived by the workers as lame, boring, or stupid. Safety presentations and displays, therefore, can be an ideal arena for employee participation. The displays don't have to be professional looking or even spelled correctly to get their important messages across. The safety messages put up by workers are more powerful than any you can purchase because they have true ownership encapsulated within them. They speak a common language between the workers. Worker participation in safety presentations, such as in workplace meetings, assures that line management doesn't forget to include a safety and health subject, builds employee knowledge by his or her preparation for the presentation, builds communication skills in workers, and displays leadership potential. The other workers also listen to safety presentations given by their fellow workers more attentively. Whether they try to challenge the presenter or listen to anything new that they can offer, this attentiveness is powerful in increasing safety and health awareness in a work team and building more knowledge about the safety and health information presented.

PARTICIPATION: WHAT DOES IT BUY YOU?

It's not a mystery why these four areas of employee responsibility in safety and health are successful. They have been successful for three reasons. First, placing them across the entire organization is much more efficient. Second, it counters the "Omnipresence Confounder" of the classical approach. And, third, none of the skills require extensive training and retraining.

Unfortunately, none of us live in a perfect workworld. Every workplace has its positive aspects *and* negative aspects. These aspects can notably affect the opportunities for employee participation in safety and health responsibilities. What are the negative impactors to employee participation defined by? Limitations are defined by the organization's cultural beliefs and barriers, by management's beliefs and barriers, and by the level of training. Culture has a lot to say about how much responsibility, authority, autonomy, and level of participation workers are allowed. It's a classical "Old World" versus "New World" difference in the organization's culture. And, as we discussed earlier, cultural changes take time and many other aspects of change management to be successful. The beliefs and barriers placed by management can be organizational in nature or area in particular. If it is organizational in nature, it is almost always driven by the beliefs and barriers of the top person in the organization. If it is in a particular area, then individual management determines the limitations. The level of training can also be a significant limiter of participation. As we discussed earlier in this chapter, you cannot expect participation from workers if they are not adequately trained and given the skills to assume those responsibilities.

One other significant factor that can limit worker participation in safety and health. These are limitations imposed by labor contract. Generally, labor contracts are very detailed concerning safety and health issues and establish a hierarchy whereby safety and health issues are discussed and resolved. This does not mean that if you have a labor contract with your workers, you cannot increase the amount of safety and health participation of your employees. Often the increased participation will be welcomed. It will just require some additional discussions, planning, negotiating, time, and training to accomplish. A cultural change may even be necessary in some organizations.

It's indeed easy to say that total quality safety and health programs use a high level of employee participation. As you can see from the many facets of participation and the factors that limit the amount of participation, implementation isn't that easy. Usually increasing employee participation in safety and health activities, however, begins with us, the safety and health practitioners. We place the barriers by our *own* beliefs, protectionism, and paradigms. How do you know when this is the case? Listen for clues in what is said. Listen for statements like—"We've always done it that way." "That won't work." "The union will never allow it."

"You can't trust them." "My management will never listen to an idea like that." What is really being said is, "I'm afraid to try." Or, "I'm afraid to spread my responsibilities and knowledge to others because then they won't need me anymore." Employee participation is too powerful to sidestep. It doesn't pay an organization back by mere additive results. It does so by exponential proportions.

Want a frightening thought? What should our purpose as safety and health practitioners really be? From a total quality perspective and from that of our initial mission, our purpose should be to drive safety and health and the knowledge of it and its principles so deeply into the line structure and everyone within the organization that we eliminate the need for our position! Do away with our own jobs? Before immediately rejecting this thought, let me ask two questions. First, why did the field of safety and health get started in the first place—was it to improve the safety and health of workers or was it to develop a secure and powerful safety and health profession? And second, what do we as safety and health practitioners do or know that is so sacred or important that it cannot be transplanted into the workforce? You see, total quality is about improvements. One of the most significant areas where we can improve *any* workplace is to eliminate wastes. If we can accomplish our mission and eliminate the need for specialized services, whether those services include those performed by quality inspectors, material handlers, industrial engineers, or safety and health practitioners, it needs to be done!

8

PROACTIVE VS. REACTIVE SAFETY MANAGEMENT

"Tom," I began after finally contacting a fellow safety professional by phone. "This is Dave Pierce. I've been trying to get you for over a week now. You're a very difficult man to reach." He responded, "Yeah, I know. Things are always crazy around here. It seems that I'm always running, trying to catch up and get everything done. It's exhausting!" "Did you lose one of your staff?" I questioned. He continued, "No, it's always been this way. You know, accident investigation here, safety inspection there, employee complaint over there, management complaint in my office, OSHA inspector at my gate...fire drills! Life around here is just a series of fire drills. I never get anything accomplished."

Sounds familiar doesn't it? I talk to safety and health practitioners all the time with the same story—different "fire drills" but same story. They all seem to be overly busy, accomplishing little, extremely frustrated, and near burn-out. Why is that? First, we are our own worst enemy. We simply cannot let someone else do what we do or know what we know. The second reason is that we don't know any other way to function. It's a pattern. It has to do with our awareness and reveals a lack of tools for changing the pattern. It's masochistic, and a form of accepted suicide.

REACTIVE MANAGEMENT

What other way can we do our jobs in safety and health? Always chasing "fire drills" is called reactive management. We are always reacting to outside stimuli. Someone has an injury, needs something, has a complaint, calls you, or uses some other "trigger" mechanism. You, in turn, respond in a reactionary manner. You go and solve the problem, do the investigation, conduct the inspection, answer the question, process the paperwork, file the report, call the person back, make a decision, etc. Trigger...reaction—that's what is called reactive management. Does this sound too familiar to anyone in the safety and health profession?

Other than the noneffective use of your time, the most insidious aspect about reactive management is the perception of management at the top. Upper management perceives a person who operates totally by reactive management as disorganized, ineffective, and inefficient. So, when an organization is pursuing a reduction in force (RIF), the "safety guy" is too often on the RIF list. Are safety and health people the only ones who are perceived as nonvalue adding? Think about it. If you have ever experienced a reduction in force, you can name the other functions that share this unenviable perception. Do functions such as training, engineering, human resources, and secretarial stand out in your memory?

PROACTIVE MANAGEMENT

There *is* a better way to do business. Call it proactive management. It's what total quality is all about. It's the mode where you actually plan what you are going to do and have the opportunity and time to do so. The two key words, they are "opportunity" and "time." Sure, there are some reactive elements in the best proactive safety and health program. These reactive elements can be significantly reduced and, depending on the success of the safety and health program, can be almost totally eliminated. Total quality can provide a great many of the tools to do so.

Besides proactive management being efficient and effective, there is another extremely important reason to move toward proactive management. That reason is that upper management perceives those who operate proactively very differently. About 180 degrees different. Upper management perceives them as competent, adding value, efficient, effective, and valuable to the organization. Considering who makes the decisions concerning safety and health manpower and budget, no other reason for embracing total quality is necessary.

Much has been said about a safety and health practitioner's need to speak "business-ese" to open the communication path with upper management. Little is written, however, about what that language really is. Too often we think that it *only* involves accounting and stockholder terms. Wrong! We also try to simplify the concept and only deal with the terminology *we* use and *miss the boat!* We miss the opportunity to change upper management's perspective on safety and health. So, we change our terminology and don't take the opportunity to tune in the receivers. What's the result? We use the right language, but because of management's feelings about our worth to the organization, management doesn't hear us. The language of "total quality-ese" can also take you a long way to better communication with upper management.

HOW CAN TOTAL QUALITY HELP?

Let's begin by answering three questions. What total quality tools and concepts are used in proactive safety and health management? How do the two worlds, reactive and proactive safety and health, differ? And, how does one deal with the change from reactive to proactive safety and health management?

First question: **What total quality tools and concepts are used in proactive safety and health management?** They fall into four areas: 1) participation, 2) clarity of line function, 3) planning, and 4) control programs.

1) The participative aspects of total quality are critical for proactive management to take place. Through participation, the safety and health responsibilities are not only spread out across the organization, but because of involvement and awareness, lessen the demand for reactive

measures. More simply stated, a high level of employee participation in safety and health places less of the burden for day-to-day activities on the safety and health practitioner or group. Furthermore, reduced injuries, illnesses and incidents that are a direct benefit of participation-induced awareness and involvement result in less demand on the safety practitioner or group to react to concerns and complaints, conduct investigations, etc. A high level of employee participation is critical for a safety and health program to change from a reactive to a proactive approach.

2) The total quality concept of clear line management responsibility for safety and health is also important. One of the major problems in a reactive management system is the "knee jerk" reactions that result when line management is not doing its safety and health job. These management safety and health jobs should include responsibilities such as conducting regular safety inspections, day-to-day observing of worker safety and health, providing routine safety and health training, performing preliminary accident investigation, addressing employee safety and health concerns, conducting and coordinating regular safety meetings, and promptly correcting deficiencies. Clarity of line function is not an option in total quality. It is a critical concept in building a proactive safety and health program.

3) Planning is discussed in Chapter 10. The interrelated nature of planning and being proactive cannot be overemphasized. You simply cannot be proactive without knowing where you are going and what you need to accomplish. Through planning, you determine where your program is going and what you must accomplish.

4) The last major total quality concept that makes way for proactive management, control programs, is detailed in the next chapter. Briefly, control programs are formal working procedures or systems that assure that an important activity is done, standardized, and documented. Because it is formalized, it requires little day-to-day oversight. Being part of the working methodologies or systems, control programs work within the normal sequence of production or maintenance. So, it is self-initiating and self-sustaining. Control programs offer great advantages in that they greatly reduce the time required to oversee covered safety and health activities. Thereby, the safety and health practitioner or group is free to perform more proactive-type activities.

THE DIFFERENCES: REACTIVE *VERSUS* PROACTIVE SAFETY AND HEALTH

Let's turn to our second question. **How do the two worlds, reactive and proactive safety and health, differ?** They are like night and day, almost exact opposites. Let's look at the differences within six of the most common responsibilities and concerns of safety and health professionals: 1) identifying and correcting unsafe conditions, 2) performing accident investigations, 3) developing corrective programs based on injury statistics, 4) selecting and monitoring the use of personal protective equipment, 5) implementing administrative controls, and 6) looking at OSHA standards and other regulatory approaches.

1) Identifying and correcting unsafe conditions: In the reactive safety and health organization, this activity is almost totally the responsibility of the safety and health person or group. It falls under that fun activity that all department personnel look forward to, the safety inspection or audit. With list in hand, the safety and health practitioner swoops down on the scheduled department at the appointed time checking for tool rests on pedestal grinders that are too far from the wheel, worn electrical cords, loose or missing machine guards, lights that are out, slick places on the floor, or cluttered storage areas and shelves. The end product is another list of things that need to be fixed, which, of course, is taken out on the next scheduled safety inspection to check for completion. It's a classic "I see—I react" by creating a paper trail.

This activity is *very* different in a proactive safety and health program. In a proactive safety and health organization, 99 percent of all workplace inspections are done by the department personnel. This breaks the "see—react" sequence. If the safety and health practitioner sees an unsafe condition, he or she asks the area personnel about it. The difference is, of course, ownership. In a reactive workplace, the safety group has ownership of the problem and must force correction without authority. In the proactive total quality workplace, the area personnel and line management have ownership *and* authority. Not only is it more efficient, it is nonadversarial.

2) Performing accident investigations: In a reactive safety and health program, accident investigations are the number one "knee jerk" reaction. If someone gets injured or a safety incident occurs, the safety practitioner or group investigates the accident and completes the report. The trigger, of course, is the accident. The reaction is the investigation and report. Other than the obvious drain on the safety practitioner's time and nerves, other costs include common and lengthy delays between the accident and the investigation, not being able to get all the facts, not being able to make realistic recommendations to prevent recurrence, or just not learning about the accident in the first place. In a proactive safety program, the first line of accident investigations and reporting falls on the involved employee(s) and line management. This assures a timely investigation and report, improves the knowledge of the processes that will result in better fact determination and better recommendations to prevent recurrence, and lessens nonreporting of accidents. Again, this is a significant difference between reactive and proactive programs.

3) Developing corrective programs based on injury and illness statistics: In a reactive safety and health program, this is usually just a dream. "If I only had the time." Reactive safety and health just doesn't allow a lot of time to collect and massage the injury, illness, and incident data, nor does it provide the time to study the data and implement corrective programs. Reactionary safety and health spends 99.995 percent of the time dealing with today's and yesterday's concerns and issues. Proactive safety programs, however, have ample time to look at today with an eye toward tomorrow. Injury, illness, and incident data can provide valuable tools for planning and standardizing safety and health programs. Not being captive to "trigger–react" sequences, allows better data to be kept, manipulated, and studied. It also allows corrective actions to be implemented and coordinated.

4) Selecting and monitoring the use of personal protective equipment: In a reactionary safety program, selecting the proper personal protective equipment (PPE) is the "turf" of the safety and health person or group. It's not vastly different in a proactive safety program from an expertise vantage. There are, however, two major differences: what triggers the beginning or change in PPE, and how the PPE market is researched. In a reactive workplace, the trigger mechanism is based on the observed

need by a safety or health person. Hopefully, the need for PPE isn't triggered by injuries or illnesses that are discovered at the medical department. Because safety and health is a line responsibility in a proactive safety organization, the trigger mechanism for selecting PPE is usually a question from line management or a from a worker. In more subtle cases, another important trigger is from injury, incident, and illness trends.

Market and vendor research is another one of those "anchors" in a reactive program that drags down the safety and health practitioner's time. In a proactive program, the market or vendor research is done by the area management and workers. After all, who is more aware of the PPE needs and restrictions? Obviously, the passing of this responsibility requires training to build PPE knowledge into the line structure *and* using simple control programs to monitor needs, progress, and changes.

5) Implementing administrative controls: In a reactive safety program, administrative controls are used only if there is a regulatory requirement to do so. Without line responsibility for safety and health issues, the common response from line management is always, "Do we have to?" In other words, "Does the law require it?" The use of administrative controls in instances where they will just simply work better or protect better is nearly impossible. This happens for two reasons. First, there is no line ownership for safety or corrective actions. As a result, second, the safety and health person or group simply doesn't have the time or authority to get a program off the ground and keep it running. That's the real rub. In a reactive program where the safety and health responsibility rests on the shoulders of the safety person or group, implementing control programs that depend entirely upon the regular and routine action and checks-and-balances within the line function is nearly impossible. At best, implementation of administrative control requires too much time and personal attention on the part of the safety and health practitioner. If the availability of time wanes because of other reactive priorities, the administrative control program goes out the window.

In a proactive safety program, however, the focus is toward the future. The safety and health responsibilities are also clearly placed within the line structure. It's a different world. Administrative controls become easier to use because, first, the safety and health person or group has more time to plan, sell, and nurture the program. Second, because safety and health is already an accepted responsibility of the line structure, administrative controls are easily incorporated into the daily mode of operation based on ease of managing efforts and protecting workers efficiently. And third, in a proactive safety program where planning, rather than regulations, form the basis of programs, nonregulatory-driven administrative controls become common and easy to use. Having line management committed to the administrative control also requires minimal oversight by the safety and health person or group. In a proactive safety and health program, implementation of control programs, even those that go beyond regulatory standards, is not only possible, it is commonplace.

6) Looking at OSHA standards and other regulatory approaches: In a reactive workplace, OSHA standards and other regulatory "have to's" become the drivers of worker safety and health. In such a world, it is common to hear questions from upper management such as, "Is it required to do so?" Or, "Does the law say that I have to do it?" These statements, of course, lead the way to the other more destructive questions such as, "Is there any way that we *don't* have to comply?" Or, "What's the chance that they will catch us in noncompliance and what would the penalties be?" This is *no* way to run a safety and health program! In a proactive safety program, safety and health are driven by statistically proven need or by efforts to protect worker safety and health. Sure, this sounds like a "motherhood" statement, but in a proactive workplace, the triggers are *very* different. The focus is different. The placement of safety and health responsibilities is different. The whole picture is different! Regulatory requirements become the "minimal acceptable level of performance."

Obviously, there are extensive differences between what can and does happen in a proactive safety program versus a reactive one. To summarize at this point, reactive safety and health programs waste time, are manpower-inefficient, cause low morale, usually require more people to do the job, waste money and resources, are adversarial, limit communication and cooperation, limit what is accomplished to improve worker safety and health, and are perceived as nonvalue adding by the line structure. These are *all* ugly and destructive to our mission!

TRANSITION TO PROACTIVE MANAGEMENT

Let's look at our third question: **How does one deal with the change from reactive to proactive safety and health management?** Transition is such a critical time. It is the period that sees the greatest confusion, the greatest potential for loss of direction, and the highest risk of failure. To be successful, change requires that a few things be done and done well. First, there must be a very clear vision of where the program is going, why it is headed in that direction, and what the advantages will be once it gets there. Second, key champions and sponsors of the change must continually expound the vision, progress, and advantages during the transition period. Third, focus on the change must be maintained in the change agents. They cannot be left out there to slog on feeling isolated or alone. And fourth, feedback on where the effort is succeeding and how much farther the effort must go is critical. It can be extremely frustrating and demoralizing to never know the score of the game. "No news is good news" does not hold true in change. In a transition period, "No news means that we are losing!" Constant feedback provides the score. And knowing whether or not you are winning is critical to a change's success.

THE ONLY WAY TO GO

Common sense should tell us that having a proactive safety and health program is the only way to go. Aside from comparing proactive programs with reactive ones, there are some key bases and focuses of a proactive safety and health program that set these programs apart from others and make them so successful at protecting workers. What are they? First, proactive safety and health programs focus on the "People Component" of safety and health, not the "Object Component." Unsafe conditions are *not* the major cause of injuries or illnesses. It's the inaction or unsafe actions of people that cause injuries. Whether unsafe acts occurred intentionally, by mistake, or by omission, they are the key to worker safety and health. Unsafe conditions are almost always the manifestations or results of unsafe actions. This is a key reason participation in a proactive safety program is so critical. Since unsafe actions can last only seconds, *all* should be observing and correcting unsafe actions in a continual effort to be effective. In fact, the higher the line structure participation in safety and health, the better worker safety and health will be.

Second, proactive safety programs use self-evaluations and benchmarking. Because they focus on today and plan for tomorrow, evaluating improvements and benchmarking against other programs become valuable tools for continual improvement. They remove the "let's try to see if this works" hit-or-miss approach to improvement. Program evaluations and benchmarking allow a program to know that it is getting better, not merely to think that it is. It also removes the negative "knee jerk" problems of reactive programs when, for some reason, the injury or illness rates jump.

Third, planning becomes a much more valuable tool in a proactive safety and health program. Tools such as process hazard analysis (PHA) are used for identifying hazards *before* they become part of a process or system. Other proactive planning tools can also become more valuable and productive to a safety and health program.

And fourth, because the safety and health program is not driven by the regulatory requirements (the wicked witch of the east, OSHA), it opens the door for the prospect of partnership in worker safety and health. A vastly greater percentage of proactive safety and health programs participate in OSHA's Voluntary Protection Program (VPP) and achieve STAR status than do reactive safety and health programs. The OSHA VPP also becomes a total quality program evaluation and benchmarking effort.

In summary, proactive safety and health programs increase the efficiency of resources and time spent by all, increase the efficient use of safety and health manpower, generate excitement and motivation, and can document real value-adding results to the organization.

Before we immerse our heads completely in the total quality cloud, however, there is also one painful reality. There will *always* be a reactive part to a total quality safety and a health program. It's the nature of our business. An accident occurs, the safety people hit the floor running. Someone finds a significant or tricky safety or health problem and the safety and health practitioner or group drops what it is doing to help its customer, the line organization. An illness is discovered, the health people focus their attention on it causes and prevention. The advantage provided through a total quality safety and health program, therefore, is that this reactive part is minimized, greatly minimized. The advantage is obvious. Total quality programs work toward continually improving line participation in safety and health. This further reduces the reactive activities and the time associated with them. Only through challenging our safety and health paradigms and the paradigms of traditional management and workers can you achieve continual improvement. It is not the answers of yesterday or today that will guide our path into tomorrow. Those new answers will come from the edge of what we know today and be seen only if we actively look for them and, once found, only if we are receptive to them.

9

CONTROL PROGRAMS

If there is one single issue that opens wide voids between people and groups or turns people against each other, it is the inability to influence the actions of others. It's the major issue that divides citizens from their governments. It's the number one issue that separates parents from their children. It's *the* issue that drives people from their respective churches and religions. It's the issue that continually inflames the United Nations. It's the issue that makes departments of the government, at all levels, not cooperate or communicate with other departments. It's the issue that puts politicians and business people at odds with their peers. It's the major issue that brings about divorce and divides families. It's the issue that motivates and frustrates special interest groups and organizations. It's the issue that provides the power behind movements such as the environmental movement. And, it's the number one issue that frustrates safety and health practitioners.

Think about it. How many times have you heard or have *you* said statements like, "That kid won't do what I want him (or her) to do." "My boss won't listen to me." "Isn't the government accountable to anyone?" "That's a stupid law." "He (or she) will have to learn the hard way because they won't listen to me." "You can't work with him (or her)." "He (or she) will just disrupt everything if they get involved." These all too common statements and questions emphasize the frustration, anger, helplessness, defensiveness, and resignation that always accompany not being able to influence the actions of others or to control the outcome of situations.

Are we all control freaks? Do we all seek to have power over everyone else? No, from the beginning of the species, we have been "helpers," not "controllers." That's why we develop social orders. As helpers, it is inherent in each of us to be team players. Sure, we like to succeed individually, but we have this deep inner need to contribute to a group or team effort. That's why we form governments, societies, companies, associations, teams, and families. With very few exceptions, we need to belong to or have identity with a group or team. It's human nature.

What does this basic human need, belonging to a group and not being able to influence outcomes or the actions of others have to do with each other? Aren't they really opposites? Isn't one a hugger-and-kisser trait and the other a trait of a control freak? Isn't that what this chapter is all about, control? Do these questions seem confusing? Contradictory? They really aren't. Let me explain. First, why do we get involved in groups? Is it to control what is going on? Not usually. We get involved to help the group's effort. Why? Because we usually believe in what the group is doing or stands for. And it is through our involvement that we hope to influence the eventual outcomes or the actions of that group. As a result, we advance what we believe in.

Second, the title of this chapter, Control Programs, is a misnomer in our society. In this chapter, the word "control" is used in the literal sense only, "to check or verify."[8] It is not used the way that society commonly uses it, "to determine the result by overseeing and dominating it." Therefore, in this chapter, control is a "soft" word, not a "hard" word. For total quality, it's important to use the word in the "soft" sense and recognize it for what it is.

Therefore, control programs in a total quality safety and health program are formal systems that check and verify that important safety and health activities and functions or related activities are done the same way each time, dependably, without requiring continual oversight. With

---

[8] Webster's New World Dictionary, 2nd College Edition

the total quality focus of formalizing the responsibilities for worker safety and health into the line structure, you can see how important control programs are to a successful safety and health program.

For what purposes can we actually use control programs in a total quality safety and health program? First, control programs combat one of the greatest frustrations of a traditional, reactive safety and health program—the "Omnipresence Confounder." We've visited this one a few times already. Control programs allow things to happen, dependably, without requiring our personal attention. Control programs can be a total quality safety and health program's best compatriot. Second, control programs provide valuable procedural benefits and documentation. Not only do they provide compliance without presence, they assure that 1) the critical functions or activities are done, 2) a proper sequence of events or steps is maintained, and 3) some verification of action or completion is kept.

What kind of control programs are used in a total quality safety and health program? Generally, control programs fall into two major groups: hardcopy programs and electronic programs. Hardcopy programs are written procedures or administrative functions that regulate activities. Electronic programs automatically regulate a function within the electronic entry or control for a piece of equipment or process.

HARDCOPY CONTROL PROGRAMS

There are many types of hardcopy control programs. The number of different applications, sequence of control, and interrelationships between controls is only restricted by the imagination of those who put the control program(s) in place. In areas where there is a need for only simple control, minimal hazard exists, or compliance documentation is a nicety but not necessarily required, single-layer control programs are used. However, in areas with significant hazard, extreme complexity, or an extreme need for compliance documentation, more complex or many levels of control are used. The most common hardcopy control programs used include checklists, verifications, key control, authorizations, certificates or records, and procedures.

Checklists are just that. They are sequential punch lists where an operator starts at the top and proceeds to the bottom, checking or completing all items until finished. Good examples of checklists are safety permits and inspection checklists. Safety permits include hot work permits, confined space entry permits, safe work permits, and excavation permits. Inspection checklists are routinely used before operating critical safety equipment such as forklifts, overhead cranes, mobile cranes, portable grinders, etc. Inspection checklists also are used to guide line management and workers through area safety inspections or accident investigations. A good step-by-step accident report form is an example of a checklist.

Verifications require that someone other than the person doing the job or activity being controlled sign-off or verify a task or function. This can be part of a safety permit, an inspection checklist, a compliance procedure, an equipment operation card, a check-out or equipment loan-out form, etc. Obviously, to make a verification control program work effectively, you need a well defined line of responsibility. Because of that, there are two requirements for control programs that use verifications. First, there needs to be enough people who are trained and responsible to make the verifications. Second, there needs to be ready access to those trained and responsible people. Verification has to be easy *and* convenient to work. If verification is too difficult, it simply will not be done.

Control of key check-out and issuance is one of the oldest forms of administrative control, and because of that, merits little discussion here. Most of us are very familiar with this type of control program. It uses simple issuance of keys or combinations to locks to control access.

Use of authorizations is usually a hierarchal control program. One important question must be asked. "At what level is authorization warranted?" The levels of authorization in budget matters are a familiar form of control by authorization. In safety and health, however, the level of authorization depends on a few factors. How significant is the hazard? What is the probability that, if uncontrolled, an unauthorized access or operation will occur? How convenient must it be to be sure that they get the proper authorizations? How often is authorization needed during normal and off-hours operation? And, is there significant potential for seeking authorization being compromised by production demands? The answers to these questions determine the level of authorization *and* the convenience issues. Failing to ask them dooms the success of authorization-type control programs.

A control program that uses return of signed certificates or records is not the best, because review of compliance is always after-the-fact. In a highly structured workplace, however, those where "verbatim compliance" is the law and not the wish, these control programs are very effective. In organizations that are highly entrepreneurial or where great freedom exists, use of this type of control program can be more frustrating than productive.

The last common hardcopy control program is one governed primarily by procedure. One significant area of procedural control is in the management of change. Changes in the workplace happen too often and in too many places for control to be handled in any other fashion. Changes in process, facilities, equipment, etc., are best controlled by procedure if there is a "this is the only way" philosophy. However, as with control programs that use authorizations, in highly entrepreneurial, "anything goes" workplaces, these types of control programs are more miss than hit.

ELECTRONIC CONTROL PROGRAMS

Electronic control programs are becoming the control method of choice for two reasons. First, electronic operation and oversight, or expert systems, are rapidly expanding in business. And second, electronic control programs provide a near fool-proof go-or-no-go system. These electronic control programs are expanding as technology and our uses of it expand. Today, however, they most often fall into three general groups: computerized or numalogic control, sequence systems, and electronic access protection.

With the expansion of computerized process controls and numalogic control systems, electronic control programs have become much easier and nearly error-proof. They are, however, driven by the economics of the process or facility and by the size and complexity of the process itself. So, you don't see many computerized control systems installed on pedestal grinders, but they are commonplace on a multimillion dollar chemical process. Other than the obvious advantage of built-in safety control features, computerized and numalogic control has another co-benefit. Because of the costs and complexity of these control systems, operators receive extensive training on the processes and operation of them. The higher level of training has also included a much better knowledge of the safety and health aspects of the process and the built-in safeguards. Both features have and will continue to provide benefits.

Sequence systems are built into computerized or numalogic controls. However, they can also be very simple in electronic design and inexpensive to install. Herein lies a two-edged sword. Because they are usually simpler, they are more susceptible to tampering by the "creative" employee. This is not to say that they are not valuable. In simpler, less complex systems, or those that bear less risk if uncontrolled, sequence control systems can be very helpful.

The last group of electronic control programs are those that protect or control access. Electronic access protection is provided in several different ways. The most common today are passwords, key switches, and badge readers. Passwords are access codes that allow a person to log-in to a computer or computerized control system. The pros and cons of using passwords include a potentially high degree of secrecy and ease of changing. The cons, however, are extensive. Cons include ease of forgetting the password, the creative ways we use to not forget the password, and difficulties of keeping passwords secret when more than one person has access to them. If operators choose personal passwords, they can be overly simple and easily guessed. And, with the increased use of modems on computers and computerized control systems, uncontrollable access by unscrupulous and creative computer "hackers" is a problem. Despite these drawbacks, passwords, due to their ease of use, are and will continue to be an important form of electronic access protection.

Key switches are a different issue. Because it relies on a physical control mechanism (keys) to control access to an electronic system, keys have fewer drawbacks. Like any key issuance program, however, the major disadvantage is that it depends highly upon the integrity and effectiveness of the administrative control program. Use of badge readers as forms of electronic access protection is gaining rapidly as technology expands. Of course, access control that uses personal identification badges incurs the risk that if this badge is lost, unauthorized use by someone else is possible. This usually has to be an intentional act, though, and is rarely done by mistake.

AN UNDISCOVERED RESOURCE

Control programs have been an undiscovered resource in most traditional safety and health programs. Control is administratively provided with extensive time demands required to administer and oversee its application. Administrative control by the safety and health person or group, because of the "Omnipresence Confounder," lack of direct authority, and the lack of ownership in safety and health responsibilities within the line structure, has never worked effectively for very long.

In traditional safety and health programs, the phrase, "It must be approved by safety" lives but, unfortunately, is commonly ignored. In a total quality safety and health program, the phrase is different. It goes like this: "It must be approved by area management." In a total quality organization, management is responsible for worker safety and health. And, only within a correctly aligned responsibility structure can control programs work. This allows magic to happen. The safety and health practitioner or group can then focus on managing the exceptions and *not* the normal situation.

Managing the exceptions, first, requires a whole lot less time. And second, managing the exceptions offers a clear picture of the inefficiencies within the control program. Having a clear focus on the problems allows a choice—improve the system or live with the problem. Problem-solving tools also become usable in such an application, whereas, before, with the problems unrecognized, no solutions were possible. Everything weaves together in a total quality program. True control becomes possible.

10

VISION, MISSION, AND PLANNING

Three words are completely foreign, without meaning, to 95 percent of all safety and health practitioners and managers of safety and health programs. They are vision, mission, and planning. An experience brought this painful reality into clear and undeniable focus for me. In a recent presentation on program management to about 300 safety and health practitioners, I responded to a question from the audience. "When you make your plan for the year, how do you know what to plan for?" We had a serious communication problem—we obviously weren't coming at the concept of planning from the same direction. We also didn't share a common language (classical "safety-ese" versus "total quality-ese"). Nevertheless, I tried to provide an answer as concisely but understandably as possible. "The annual plan is merely a portion of your strategic plan. Do you understand the strategic planning process?" He gave a muted, polite nod in response. "Well, a strategic plan is developed to meet your specific mission. Do you understand what I am saying when I refer to an organization's mission?" Again, he nodded politely. However, this time a faraway look was starting to appear in his eyes. "Not wanting to get too complex about this, your program's mission is derived to support your program's vision." At this point, he looked like he was hit with a brick. "Does any of this make sense to you?" I asked. "Uh...yeah, I think so...uh...no...well...I *am* a little confused about your plan still...how do you know what to plan for?" Stepping back and addressing the full audience I asked, "Does anyone know what I am talking about? Is anyone involved or familiar with strategic planning?" It was like asking for volunteers to crawl headfirst into a well. Turning to the session monitor I commented, "You'd better invite someone to come to talk about strategic planning and total quality."

This is not an isolated experience. Most people have little knowledge of vision, mission, and planning, and their interrelated nature. That should be no surprise. Safety and health programs have historically been reactive-oriented, not proactive and future-oriented. "I'm too busy just trying to keep my head above water," a safety and health professional friend of mine said. "I exercise planning in my day planner, but it is seldom more than a week out. And mostly, it's the meetings I have to attend."

If safety and health programs have existed using this "by the seat of your pants" approach this long and the profession still exists, what's wrong with this approach? If you accept the idea that traditional safety and health management is the best way, and that business concepts need not change, you might then conclude that this approach is okay. However, you couldn't possibly believe either of these thoughts to be true or you wouldn't be reading this book! The fact is, there are *always* better ways to do things, including managing safety and health programs. This is the foundation of continuous improvement. The second fact is that business *is* changing and changing rapidly. These changes are coming from a new way of thinking. Businesses of tomorrow will look little like yesterday's, or even today's businesses. So, change we must, if our mission is to be successful or survive in this new business world.

VISION

What drives change? Does change occur out of need or by accident? Positive change must be planned and directed. Positive change must begin with a vision of the future. A vision is the horizon, the catalyst and reason for that change. Many years ago, Chrysler was on the verge of bank-ruptcy. It wasn't struggling. It was all but dead. Out of nowhere came a visionary, Lee Iacocca. He saw Chrysler's future. His vision literally turned that corporation around. History recorded the success brought about by his vision. Think for a minute. What would have happened to Chrysler if the vision had been "business as usual"? Do you remember American Motors? How about Studebaker? You can see very easily how important and basic a vision of the future is to change.

Not all visions are great, however. Nor are all accepted by those who steer programs, organizations, companies, or countries. Steven Jobs had a vision about computing, but he had to form his own business, Apple, before his vision was accepted. Great visions challenge classical paradigms. That's why they are difficult to see and accept. Great visions come from the edge, challenge the status quo, and think in ways that are foreign to us. Embracing total quality is, in a way, a vision of the future. It is definitely not the way we have conducted business, or at least the business of safety and health programs in the past. And, like other visions from the edge of our paradigms, it suffers the same problems—the inability to be seen clearly and objectively by others. It is easy to slip back into the misinterpretations and misunderstandings that surround total quality. "Isn't that just a manufacturing concept?" "How is statistical process control going to help your safety and health program? After all, isn't total quality just statistical process control?" "Management by objectives is dead!" "We're just too busy getting the daily work done to think about total quality!" "The boss will never go for a change like that!"

Visions usually encompass many aspects of a business, organization, or program. For example, a change in vision for a safety and health program may focus on acceptable levels of exposures, the use of personal protective equipment, what "compliance" means, or increasing employee participation and responsibilities. Visionaries do not ask "why?" They ask "why not?" They are positive-oriented and never negative. Many years ago, I took control of a safety and health program that was marginal at best. The lost time injury rate was well over 2 percent. It was common and expected to lose more than 300 workdays each year due to injuries in a plant with fewer than 500 employees. The recordable injury rate was near 15 (15 out of 100 workers), and the total injury rate was near 50 (1 out of every 2 workers). Management hired me with one purpose in mind—bring the numbers down.

My vision was different, however. I wanted the safety and health program to be the best in the state, the best in our operating division, a leader in the corporation. Everyone to whom I told my vision thought me a little strange. I received the same polite but disbelieving response that a four-year-old would receive turning cartwheels on the lawn with aspirations to win Olympic gold. Six years later, however, the plant became the first facility in our state, in our OSHA region, and in our corporation to achieve STAR recognition under the OSHA Voluntary Protection Program (VPP). You should have heard the "boo-birds" when we announced that we were going to submit a VPP application to the "enemy," OSHA. "You can't do that!" "That's stupid!" "The corporation can't support your actions." "You must be masochistic *and* stupid!" Visions that come from the edge of others' paradigms of how things are supposed to be or thought are *not* easily accepted. Thomas Kuhn told us that. Joel Barker emphasized it too. It is also a fact when it comes to visions within safety and health programs.

Vision, therefore, cannot be a "hidden light." It has to become a common vision if it is to have any chance at driving change. How does a vision become common? Obviously, there are some communication and sales skills are necessary. Generally, it begins with sharing the vision or telling others about it. A word to the wise. Sharing a vision shouldn't be just, "I think we should do it this way," or "I think that our thinking should change." When sharing a vision, especially if it is notably different from the common paradigms, you also must include a statement of mutual benefit to your audience. Such as, if we do this or think this way, this will be our reward. It's an "if-then" statement and is critical if anyone is to hear you. After sharing the vision, you must sell it actively. Lee Iacocca did this in the Board Room, to his employees, and to you via television. And he continued to sell his vision year in and year out. He sold it so well that his vision became inseparable from the company and the cars the company made. Only through active sharing and selling a vision can it gain acceptance.

Does it take a Lee Iacocca to sell a vision? No, but it isn't an exercise for one who is not committed to the change or one who wears feelings on the sleeves. You must be thick skinned and *very* persistent. If you allow the "boo-birds" to turn you off, your vision and the change will go with it.

What do you want your safety and health program to become? Are you content for it to be as good (or bad) as it is today? What common ways of thinking in your organization or in your safety and health program are roadblocks to improvement? What basic change in thinking will turn your program around? Look to the edge of your Safety and Health Paradigm for the answers. Therein lies the vision for change.

MISSION

"Sure we have a mission statement. We derived it in a group meeting about two years ago," she continued. "What does it say?" I asked. "Well, let me think. It says something about...hum...I don't remember exactly." "Didn't you write it down?" I asked. "Sure! I don't remember what happened to it, though. It's probably in one of my files somewhere." "I see," I continued. "How is your mission statement driving your program and actions, then?"

Everyone has heard of mission statements, but do most of us really know their purpose or what makes a good mission statement? What is a mission statement supposed to do? It isn't a fad. It actually has a very important purpose. Simply, a mission statement tells why the company, organization, or function exists, nothing more, nothing less. It answers that basic but critical question, "Why do we exist?" Sometimes mission statements have other information such as the values by which the mission will be guided. If short and concise, these don't detract from the mission statement. The danger is in adding too many ideas or too much wording, which increases the chance that the real mission will be lost in, or modified, or contradicted by the wording.

Good mission statements are short and don't waste words. Typically, they have a few key words built into them and answer three questions: Who, How, and Why? For example, "The Health, Safety, and Environmental Group is committed to providing protection and prevention services by elimination or control of hazards to control human and financial costs." Short and sweet. Let's dissect it a bit. There are three key words: Group, services, and costs. The group defines "Who" exists because of the mission. Services refers to "How" the group will accomplish its mission. Costs tell "Why" the mission and group are important to the organization, a statement of benefit.

A mission statement tells everyone, including you, your staff, your customers, and your boss why your function is important to the organization. A mission statement is best derived by group or team because of the different perspectives and needs of each of the team members. It also builds ownership in the statement itself. A mission statement should be approved by your boss and the head person at your facility (if different). It not only gets their buy-in to your purpose but also serves as a strong reminder of why your function is important to the organization. Sales of this point are *never* wasted efforts.

A mission statement is not a "puff in the wind." It is a lasting document. So, after you've formalized your mission statement and gotten it approved, frame it and display it. It can serve as a constant reminder to all. If you bury it in the file, it and the importance of your function will be forgotten by you, by your staff, and by your boss. Each team member should display the mission statement his or her desks. Visible, mission statements provide a very clear team focus. Too many times due to daily frustrations, being busy, etc., we lose focus on why we exist. A visible mission statement can provide that valuable daily reminder. A good total quality program will always have a mission statement.

PLANNING

What is strategic planning? It is derived from the word strategy, which is defined by Webster as: "the science of planning and directing large-scale military operations, of maneuvering forces into the most advantageous position prior to actual engagement with the enemy." From this definition, it isn't difficult to come up with a good mental picture of a Board of Directors maneuvering corporate forces before engaging the competition. Very military-like in their approach, corporations and companies have used strategic planning for a long time to discover and take advantage of competitive strengths and shore up weaknesses. They plan how to not only stay in business, but how to increase profits and market share. How is this concept of strategic planning applicable to total quality safety and health programs? Let me offer a less hostile, friendlier definition of strategy as it might relate to safety and health. Paraphrased from Webster, it would go like this: the process of planning and directing a coordinated program effort, of activities and resource-building to assure compliance, improve worker safety and health, and move toward continuous improvement.

Many books on strategic planning are available; unfortunately, none speak of this process in staff or service activities. So, let me summarize the important elements of strategic planning for use in a safety and health program. Basically, a strategic plan tells what you want to accomplish or become in a long-range time frame, what you have to or want to accomplish in the medium-range, and what you need to accomplish this year to support compliance and your medium-range and long-range plans. A strategic plan isn't day-to-day, week-to-week, month-to-month, or year-to-year. It is usually five to ten years ahead in time. Obviously, a vision of the future and a mission are important to the strategic planning process. Strategic planning cannot be effective unless a vision points the way to success and the planning supports the mission.

Strategic planning usually has at least three periods: long-range (five to ten years into the future), medium-range (bridges the gap between the long-range plan and the short-term or annual plan), and the short-term or annual plan (one-year plan). The long-range plan positions a program to meet the vision of the future. It is usually not an extensive list, but a short list of three to five items with desired attainment dates. Long-range planning items might include being able to submit a VPP application, making a quantum change in injury or illness rates, or having all program data and records or program coordination computerized. Medium-range plans look at more specific items that need to be accomplished to meet your long-range goals. These might include sequentially lowering the injury or illness rates, computerizing a percentage of all program records, or completing a safety and environmental policy manual. Medium-range plans have greater detail and have more items than long-range plans. They are inseparable from the long-range plan. Short-term or annual plans are different. They not only contain items that must be accomplished to meet both long- and medium-range milestones, but they also contain those things that need to be accomplished during the year. These must-do or need-to-do items may include reports, filings, permits, inspections, audits, training, formal communications, etc. The annual items that are tied to your other plans would include targets such as making a particular recordable injury rate, or writing a specific safety and health policy for the manual.

The most important aspect of strategic planning is the process itself. The planning process is usually done once each year, at the beginning of the year or at the end of the previous year. It begins with an affirmation and, if necessary as driven by a new vision of the future, a change in the long-range plan. Second, the medium-range plan from the previous year is evaluated, revised, added to, etc. Third, the last annual plan is reviewed and revised in two parts: 1) the items that move toward the medium-range and long-range items, and 2) the need-to-do or must-do items that will be required during the upcoming year.

SUCCESSFUL PLANNING

Some important aspects of these plans that make them more successful are listed below:

- ► Assign responsibilities and dates
- ► Prioritize the Short-Term Plan
- ► Get the Plan "approved"
- ► Display the Plan
- ► Review the Plan regularly
- ► Make the Plan part of your Performance Management System
- ► Plan by team

First, assign responsible people and dates to each annual plan item. It's the only way things get done. Second, because the strategic planning process is so dynamic and usually looks at ideal conditions, break your annual or short-term plan into "Have-to-be-done," "Need-to-be-done," and "Would-like-to-get-done" divisions. Work by the "100/80/50-Rule," accomplishing 100 percent of the "Have-to-be-done" items, 80 percent of the "Need-to-be-done" items, and 50 percent of the "Would-like-to-get-done" items. Using this rule avoids frustration and discouragement. Third, get your plans approved by your boss and the head person in your organization. It is critical to the success and support for your program to do so. It is also a good PR tool to use. Fourth, post your plans where *everyone* can read them, like on a total quality or public-access bulletin board. It lets everyone know where your program is going, what is being done, and helps to keep you focused. Fifth, regularly review the plan against what you've accomplished. Once or twice a year is *not* enough. At a minimum, it should be reviewed quarterly. Monthly is better. Sixth, make your strategic plan part of the performance management system that rates you and your staff. This adds the reward if accomplished and correction if not. And seventh, just like writing a mission plan, the planning process should be done by team. Allowing team members to identify their own "to-do" list and set their own completion dates brings ownership, is self-managing, and highly motivational.

When talking about planning, a saying comes to mind. Originally, it was not intended to focus on the safety and health field. If it had, it would have been written by a "punster." I took the liberty of paraphrasing the saying to increase the intended impact. Revised, the saying goes like this. "If you do not work by plan, you will most assuredly fail by accident."

Vision, mission, and planning aren't just fads. They aren't empty words to impress the boss or peers. They are important to a successful total quality program. They point the way to the future, detail why the program adds value to the organization, and set up a step-by-step plan for accomplishing success and continual improvement. If you choose to place total quality into your safety and health program, these are not optional. They are basic building blocks, foundations if you wish, to your program. Without them, your program cannot move ahead with total quality. They are essential!

11

GOALS AND OBJECTIVES

As children, one of the questions we are asked most often is, "What do you want to be when you grow up?" Like most kids, I went through many phases in my early years where what I dreamed of being changed dramatically. One time I would aspire to be President of the United States; another time, a doctor. I also pictured myself as a fire fighter, a police officer, a jet fighter pilot or a military general. What did you want to be when you were a kid? Did you become that or something else?

Throughout our lives, we have all had goals. Today some of us have many goals, some few. Some of us have lofty goals, some closer to the ground. Some of us actively look for goals and measure our progress toward them, some take little or no interest in achieving them. In fact, when you look at goals, goal setting, and progress measurement across the population, the spread is very wide. It spans from extreme goal-oriented behavior to existence almost without any goals.

Why are goals important? First, if you don't have a goal, how can you keep focused or know when you have done what you set out to do? It's the difference between moving forward and being parked. Second, goals add a lot of interest and excitement to life. Goals get our blood moving and keep us interested instead of bored. Third, achieving goals makes us feel good about ourselves. Otherwise, we just exist. And fourth, goals make us more interesting to be around. Why? Because, we talk about them and we get enthusiastic about them. Whether the goal is our planned vacation, retirement, family plans, building a house, or anything else, we are more interesting to those around us because we are more focused and enthusiastic about life.

Survey after survey gives us the following economic breakdown of people in America. Three percent of the people are financially independent. Ten percent are living comfortably. Sixty percent are just making it, defined as living within plus or minus ten percent of their income. And twenty-seven percent of the people need some kind of public aid to live. What's amazing about this distribution is that it's been stable for years and years, changing very little if any.

Financial experts go farther though. They say that if you spread out all the personal wealth (or lack of it) evenly across everyone in America that within seven and a half years, the same 3%-10%-60%-27% would reappear as if it hadn't left. Why is that? Those financial experts believe that the largest single reason for this is the existence of goals. The three percent that are financially independent have many written goals that are reviewed regularly and a plan that exists with milestones and measurements toward accomplishing those goals. The twenty-seven percent who need some kind of financial aid to get by rarely have any goals, and if they do they are short-term and usually focused on simple finances. As an example used by Coonradt, "to cover the check they just left at the grocery store."

If goals are that important to our success, then why don't we all use them? Is it unnatural? After all, when we were young, we had goals about what we wanted to be, didn't we? Psychologists tell us that goal setting is indeed natural to us. Business experts expand on that thought by adding that goals and the goal setting and tracking mechanisms are vastly different in people, and as a result, also in organizations. History would certainly support this idea in that there are definitely people and organizations that are successful and those that are not. There are, of course, wide differences in perspectives about the importance of goals and goal-setting skills in people and organizations. Why is this? From our formative beginnings, goal setting *is* natural to us. The skills to set and use goals effectively, however, are not. That's where the disconnect occurs.

GOALS IN SAFETY AND HEALTH

Are goals important to a safety and health program? More than you may think! Here's where a major break exists between traditional safety and health programs and total quality programs. Traditional programs make few goals, and when they do, they are usually poor ones or not measurable. Total quality programs have many goals and measurement systems. Is this difference in approach harmful to safety and health programs? It is one of the major reasons that in traditional safety and health programs there exist frustration, burn-out, low energy, poor attitudes, and low motivation.

Let me ask a question. What are the top, most commonly found goals in *any* safety and health program? Look at any program. You'll find the all too common ones are lost-time injury rate, lost workdays, recordable injury rate, and total injury rates. The "Big 4!" "So?" You ask. "What's wrong with them. If everyone has them, how can they be bad? If they're so common, how can they be wrong?" Let me answer these questions by going back to the first of this chapter. Did you ever want to be President of the United States? Did you ever set a high career goal for yourself? How did you do? Like most kids you probably became disenchanted with your lofty career goal or became discouraged and changed your career goal to a much more achievable one, didn't you? Why did that happen to most of us? Focus on this next point because it lies at the crux of this goal issue. We changed our goals because we had little or no control over achieving them. It didn't make any sense. Chasing a career goal that was unachievable, more a function of opportunity and someone else "opening the door," was rejected, and we changed our course. That's a basic rule concerning good goals. They must be under your control.

Now, looking at that analogy, let me return to those questions about the "Big 4" goals in safety and health programs. What is similar between a kid's career goal to be President of the United States and a safety and health program's goal to achieve a recordable injury rate of 0.0? Neither are under the control of those who set the goals!

Other problems concerning traditional safety and health goals are that they are amorphous. Common safety and health goals that fall into this category include to achieve compliance and to provide a safe environment. What do those two words, compliance and safe mean? With the ever-changing regulatory environment, the existence of countless codes and standards, and the blurring line between legal and liability requirements, the term compliance is nondefinable *and* nonachievable. Also, there is no such thing as a safe environment. Webster defines safe as the absence of harm or risk. Have you ever heard of *any* environment that is free from harm or risk? Of course not! A safe environment is a concept only in the minds of extreme idealists. It just doesn't exist!

What is the problem with this traditional approach to safety and health goals? Consider that the traditional goals are amorphous and subjective, not measurable, depend on outside evaluation, and simply aren't achievable. Furthermore, as we have learned the hard way in business, the "Big 4," are subject to recording games that have resulted in many multimillion dollar fines by OSHA. They are also not indicators of a good or bad safety and health program and, at best, are only reactive indicators. In all, these traditional safety and health goals are not understood by upper management, difficult to put a value on, and result in frustration, burn-out, and high activity with low results. These represent an impressive list of problems.

THE RULES OF GOAL SETTING

Developing good goals and objectives doesn't involve magic. It is simply a skill that can be mastered. If followed, some general rules covering how goals are stated and how they are best worded assure good goals and objectives. First, **goals must be specific.** Goals like, "We want to be successful," or "We want to lower our recordable injury rate," aren't any good because no one knows when to celebrate the accomplishment. After all, if a goal is so "wishy-washy" that no one knows when it is accomplished, it really isn't a goal at all. It's only a direction. A specifically worded goal might read, "I want to complete five department safety and health audits each month," or "Complete writing an Ergonomics Policy by October 4th."

Second, **goals must be positive.** The human mind is constructed so that it has a hard time distinguishing the difference between similarly worded negative and positive goals. It also has a hard time mentally seeing negative goals. For these reasons, personal goals such as, "I want to lose weight," or "I want to quit smoking," have always had poor chances of success. Our minds become confused. As a result, we are surprised when suddenly we become more hungry or when smoking becomes an irresistible urge. Besides, we have a social confounder, baggage from our upbringing. Our parents never raised us to be losers or quitters, did they?

Third, inasmuch as possible, **goals need to be under your control.** For example, having a goal to make your husband quit smoking cigars or your wife stop putting on makeup while driving has little chance of success. Goals that are beyond your control only invite frustration and resignation. If these are the inevitable results, why would anyone set a goal beyond his or her control? Knowing that these kinds of goals are common, this is an interesting question, isn't it? Unfortunately, this general rule is very often overlooked in our private lives, at work, and in other aspects of our lives. Want a guaranteed example? Do you have a child? What are *your* goals for him or her?

Fourth, **goals must be written and visible.** Writing the goal down and making it highly visible keeps focus on the target. Too often, goals are set but are not written down or they are written down but lost. What happens? That's right! They are quickly and easily forgotten.

And fifth, **goals must be realistic and attainable.** It's like that young child who sets a goal to become President of the United States or the second Michael Jordan. This is a two-edged sword. Goals must be realistic and attainable, but they must also be something to reach for, something that requires extra effort. If goals are too easily achieved, they provide little satisfaction and little motivation. Goals must be:

> ►Specific
> ►Positive
> ►Under your control
> ►Written and visible
> ►Realistic and obtainable

GOALS VERSUS OBJECTIVES

What's the difference between a goal and an objective? It's common to use these two terms interchangeably, but the difference between them can be used to add structure and power to the planning process. Goals are more long-range than objectives and they may or may not be quantitatively measurable. Objectives, on the other hand, are shorter-term than goals and are *always* measurable. Most often, objectives are goals that are broken into small, progressive pieces. Good examples of goals would be statements such as, "By the end of the decade, we want to be recognized by our customers as the highest quality supplier," or "We want to have a better recordable and lost-time injury rate than the corporate rates and to maintain that lead." These goals are long-range targets that are not easily measurable. Like total quality, the targets move. Objectives leading toward these goals might be, "In December, we will conduct our customer satisfaction evaluation and plan a strategy for improvement based on the evaluation's findings," or "At the end of this year, we want to make or better a recordable injury rate of 1.5." Both objectives are definitely measurable and, at most, would take one year to achieve. This allows everyone to know when it is time to "pop the cork" and when it is time to set new, more demanding objectives toward eventually reaching those long-term goals.

MAKING EFFECTIVE GOALS AND OBJECTIVES

How can we make effective goals and objectives? What tools or techniques can you use to maximize your chances of success? Here are five guidelines for setting goals and objectives. First, set your goals and objectives as a part of your planning process. Objectives that stand alone and are not an integral part of your planning process toward continuous improvement are much less effective than those that are.

Second, word your goals and objectives clearly. It's okay to be wordy if it clarifies the details of the goal or objective and eliminates the need for interpretation. Poorly worded or unclear goals and objectives have two confounders. What you are trying to accomplish is foggy to those around you and to your group. And because it is open for interpretation, along the pathway, everyone, including you, is tempted to bend or change the definition to meet new challenges. We call it rationalization. As Coonradt wrote, "When you bring your behavior up to meet your goals, it's accomplishment—when you bring your goals down to meet your behavior, it's rationalization."

Third, as was stated in the "rules" portion, goals and objectives must be realistic. For example, setting an objective of a zero injury rate is great on paper, but not realistic. The absolute best that can be done is to *meet* the objective. Because there is no such thing as a perfect safety and health program, most of the time the objective will be missed. What happens? The objective becomes a "wish list" rather than a "dedicated focus." It's like having a dream of driving a Ferrari and owning a Chevy.

Fourth, accomplishing your goals and objectives should be under your influence or control. This can be one of the frustrations within a staff function. Liaisoning to the line structure, most of the traditional safety and health objectives are under someone else's control and influence. Using some other tools like X-Matrix can notably decrease the frustrations and increase the team buy-in, but the risk for frustration and missing objectives is large. It is, however, unrealistic to only set goals and objectives that are totally within your power. Safety and health is most often a staff function and does not have direct line responsibility. For those goals and objectives that lie outside the sphere of influence, it takes a lot more coordination, coaching, leading, communicating, and nurturing to bring about success. This is a "hand-in-glove" reality. It also is an opportunity to use tools like X-matrix and communication.

Fifth, objectives must be assigned to someone. Seems rather obvious, doesn't it? Sad to say, it isn't. Often a realistic objective is not met because no one was assigned to accomplish it. The objective gets lost in everything else that's assigned.

And sixth, completion dates must be established for each objective and goal. Don't leave an objective out there with no cut-off or accountability point. That's when the "Procrastination Bug" sets in. If you plan on getting something accomplished, you *must* know when!

Guidelines for Setting Goals and Objectives:

► Make them part of your planning process
► Word your goals and objectives clearly
► Make goals and objectives realistic
► Ensure that goals and objectives are under your influence or control
► Assign objectives to someone

ACCOUNTABILITY

Accountability is a critical aspect of getting anything done. You simply must know who is responsible for the task. Saying, "okay team, let's get it done," never has worked unless someone in the team voluntarily takes the challenge. There are important parts to accountability. Dissecting the term, it has four parts: person(s), time, milestone(s), and formality. To make someone or a group accountable for a task or responsibility, you must first identify the person or team. Second, you must set or agree upon a time or date for task completion. Third, if the task will take a long time, is complex, or the person or team is on a learning curve, you need to set milestones, times or dates along the path toward completion where progress is tracked. Fourth, and very important, a formal system must be involved. The goal must be written. The objective must be written. The person(s) charged with the objective must be noted. The targeted completion dates and milestones must be written. And, for added punch, there must be a tie-in to the person's or team's performance management system because without a reward system, accountability is tough to maintain. This is an important part.

In total quality, everything is formal and interwoven: vision, mission, planning, goals, objectives, accountability, participation, training, communication, *and* rewards! If you use any total quality tools, including goals and objectives, you must tie them to the performance management system that your organization uses to measure and reward employees by their pay. The performance management system cannot be isolated from this bonding with total quality. It is the personal tie that binds the whole system together.

A friend has a challenging problem with her son. He is now almost thirty years old, still living at home, and sometimes employed. When he was struggling to stay in high school earlier in his life, there were many sometimes-heated conversations about the importance of getting a high school diploma. He was bored and unconvinced. Unfortunately, due to his lack of performance at school, he was invited by the high school to leave. He did. What were the consequences of his lack of performance? His mother bought him a motorcycle for transportation and moved him into the basement apartment where he would be less disruptive. Was his performance tied to his reward system? Did this disconnection affect his current and future performance? Is formally tying objectives and performance against those objectives to an organization's performance management system important? You bet it is!

Steps to Accountability:

► Identify the person or team
► Set a time or date for completion
► Set important milestones
► Write down everything
► Tie accomplishment to the pay system

A frustrated environmental engineer on my staff came to me the other day. He was trying to get an operating department to comply with an environmental regulation. "Every day I go by the area and they simply aren't doing what I've asked them over and over to do," he said. "I'm past frustration. What can I do?"

My approach was a simple three questions. "Is it formal? Is it visible? Is there accountability?" In this particular case, it was formal. Procedures were detailed, in place, and all operators were trained in proper procedures. It was only visible, however, by the amount of collected waste and there was no tie-in to the employees' performance management system. As a solution, the environmental engineer placed signs at the accumulation site stating the limits. The problem was also made visible to upper management in monthly compliance reports. And, talking to the area management, compliance was tied to the employees' performance management system. An amazing thing happened. The problem went away!

SCOREKEEPING (MEASUREMENT)

One of the areas most often forgotten or misapplied in managing goals and objectives is in the area of measurement, or what I call scorekeeping. Scorekeeping has two purposes. The first and most important is motivation. If you are winning, you are more apt to continue to maintain your performance if you keep score. The second purpose of scorekeeping is that if you are losing, it allows you to change your performance so you can win. If you are losing, keeping score points out the need for you to change your performance. Take sports as an example. They keep score very well! In any team sport, if the scoreboard shows a team to be winning, does it motivate them to maintain that level of performance? If the scoreboard shows that they are losing, doesn't it make them try a little harder or cause the coaching staff to try something else to improve the performance? Too often measurement is only seen as unnecessary paperwork. Not true! Measurement is critical to not only managing any function, it is critical to motivation. Measurement needn't be cumbersome or complex. It just needs to be done.

Measurement:

- ▸ Motivates
- ▸ Allows performance to be changed

How is measurement most often misapplied? The most often abused area is in what we measure. It is common to measure activities and not results. The origin of this problem goes all the way back to management by objectives (MBO). For example sales calls, documents copied, forms processed, safety inspections performed, medical examinations completed, etc., are measured. These are *not* quality indicators. If you measure safety inspections completed, what non-quality confounders are you inviting? The inspections become quicker, not longer. They become more superficial, not detailed. More haphazard, not investigative. They become more impromptu, not planned and scheduled. They get done at the end of the month, not throughout the month. They become one-personed, not team-oriented. And communications and documentation are forgotten or compromised. If, however, findings per inspection or resolved or corrected findings are measured, you measure results. Too often we are drawn to activities, "I'm busy, what else is there." And, we forget that *we are not employed to be busy. We are paid to accomplish results.*

Measure results, *not* activities

The second area where measurement is misapplied is when the scorekeeping is buried. It resides in some three-ring binder somewhere, in a file, or in the pile of papers on the boss's desk. It is not visible. Measurement systems that lack visibility quench involvement, don't motivate, are forgotten, and lose focus. A good example of this is the difference between crowd involvement in ice hockey versus figure skating. Is there a difference in crowd involvement and excitement? Think about it. Crowds watch figure skating politely and quietly. When the skater or skaters finish, the crowd applauds politely. Why? Well as Chuck Coonradt put it, "Because none of them know what they just saw." What happens when the judges post the results? The crowd begins to respond because they now know what they saw. The crowd response, however, is greatly tempered because it is after-the-fact. Were the skaters motivated during their performance by the score? How could they be?

What about ice hockey? Is there a little difference in crowd involvement and excitement? Would you say that the crowd is a little more involved in ice hockey than in figure skating? The difference can be found in the scorekeeping that is used. In ice hockey everyone watching knows the score at *all* times. Are the hockey players motivated by the score? The point is that in hockey, the scorekeeping is visible and dynamic. In figure skating, it isn't visible and only appears after the performance. Get the connection between the importance of visible measurement and involvement and motivation?

> ## Measurement must be visible

The last area where measurement is misapplied is in timeliness. This happens because scorekeeping is considered to be too much trouble. So, scorekeeping isn't kept up to date. It lags off a month, two or three. What happens? Everyone stops looking at it! It ceases to exist. Motivation falls off. Feedback ceases. Knowledge of the score is lost. Involvement is subdued. For scorekeeping to be effective, it must be kept up to date. Think of it this way, what would happen in the middle of a NBA basketball game if the scoreboard lagged behind the scoring. Instead, a sign begins flashed saying, "Update in 10 Minutes." It wouldn't happen and we wouldn't tolerate it, would we?

> ## Measurement must be kept current

Scorekeeping or measurement provides one of the most basic human needs—feedback. Humans cannot exist without feedback of some sort. It is a primal need. Feedback determines who we are, what we can do, and how we feel about ourselves. For example, kids who are constantly barraged with "You're an idiot" comments, turn out to think that they are dumb. Those who constantly hear that they are ugly or awkward, think of themselves as they are told. How many times have you heard about people who were "turned around" by a special person, running with a new crowd, or changing a job? What happens in these instances is simple. They change the feedback they receive and, so, it changes them!

My son has a friend. He's a nice guy but has always been very negative. He was known for his moodiness and would never pass up an opportunity to "put someone down." These weren't some of his more admirable qualities. We knew that his home life was anything but supportive and loving. "What can you expect?" My son would tell me. One day, he met this girl. She was one of those "sunshine people," always smiling and quiet. She was a *really* nice person. They got serious about each other, as 20-year-olds often do. A strange, but predictable thing happened. My son's friend changed, not just a little, *a lot*! Overnight, he became upbeat, excited, and stopped putting people down. It was actually a pleasure to be around him. What happened? The feedback he was receiving had changed. Consequently, so did he!

Our need for feedback is undeniable. It is so deep that it can also be destructive. For example, it is a natural subconscious reaction when positive feedback is not received to invite negative feedback instead to meet the need. For example, look at kids. They want to show you their homework, tell you what happened to them today, play catch or cards with you, show you their coloring book, show you how they can play their computer game, etc. What are they seeking? Feedback! What happens if you ignore them? They *change* their behavior to get a different kind of feedback and, thereby, satisfy their need. They start to cry, they track mud on the carpet, they leave the door open or slam it, they don't pick up their messes, they leave the light or computer on, they don't do their homework, they fight with their brothers or sisters, they stay out late at night, etc. They willingly but subconsciously invite negative feedback. They can't help it. Neither can we.

This seeking negative feedback when positive feedback is withheld. It is an important force on the job. What kind of feedback do you receive from a reactive safety and health program? Is it positive or negative? Do you feel good about your performance and motivated from the feedback you receive or are you frustrated and demoralized by it? By choosing a total quality safety and health program, you can influence the type of feedback you receive. And, like it or not, you are not isolated from the need for feedback. If you think you are, count the number of mirrors in your house or apartment!

Measurement provides feedback

There is another problem with humans and feedback. It comes from the "world owes me" monster inside each of us. The belief says that it is someone else's responsibility to provide us with the amount of needed feedback. That responsibility may fall on our spouse, our boss, or our parents. Coonradt emphasizes that the appropriate amount of feedback can only be determined by the receiver and *not* the feedback provider. Does it make any sense then that we expect someone else to guess when and how much feedback we need? Talk about openly inviting frustration and animosity. It doesn't work! It can't work! This is another reason scorekeeping is so important. Scorekeeping provides feedback. If we determine our own scorekeeping system and keep our own score, we sidestep this inner "world owes me" monster. Feedback then becomes our responsibility, as it should be. Feedback becomes dependable, not haphazard.

There is a simple correlation here. Visual, dynamic results-oriented scorekeeping provides feedback. Feedback motivates people. This is not brain surgery! It is a simple but important correlation that we *must* understand among scorekeeping, feedback, and motivation. Again, simply stated, scorekeeping provides feedback that motivates us.

BEING WINNERS

If there is one single, most important aspect that separates successful persons and programs from those who struggle or fail, it is the existence of goals and objectives. Successful persons and programs know where they are going and how they are going to get there. They measure their path to success. Unsuccessful persons and programs don't. In the words of a well-known saying, "If you don't know where you are going, any road will get you there." Total quality programs are successful because they know where they are going. Goals and objectives provide the focus and direction that drive a total quality safety and health program toward success. Measurement provides feedback and the assuredness that we are winning!

12

TRAINING, INFORMATION, AND COMMUNICATION

In a fellow safety and health practitioner's office was a rather dominant, large-lettered sign on the wall behind him. It said, "I must be a mushroom the way they keep me in the dark and feed me bullshit." It's a humorous way of looking at a very un-funny situation that most of us routinely find ourselves in. We are in a steady state of not knowing what's going on. It doesn't matter if you refer to our personal or business lives. We feel that we are constantly "in the dark." Whether it's not knowing that a credit card balance is out of control, not knowing that your daughter is on the pill, your son put a dent in the car, a new production process is being considered at work, or that your boss is considering reducing the size of your group, we all have been in the "I didn't know" position.

Reality is that, first, there is no way that we could ever know everything. Second, there are so many people that affect our lives, our jobs, our work, etc., that communicating as much as would be necessary to keep us informed as much as we would like would be impossible. We would spend all of our time communicating and getting nothing done. And third, there are some things that we just would prefer *not* to know. Sometimes, we prefer being left out of the communication pathway. Accepting these three facts, the mushroom position is as natural *and* acceptable to us as breathing.

When you begin a topic such as communication, information, and training, it is always important to begin with a reality shot such as this. Otherwise, it is too easy to get caught up in the "motherhood" components of these three words. After all, in our hearts, we all believe that everyone needs to be adequately trained, deserves complete information, and should never be left out of the communication loop. The problem with such an approach is, of course, it just isn't possible or practical. So when we talk about communication, information, and training in a total quality program, what level are we talking about? Let's start with training. It is basic to both the other terms and sets the foundation for discussing communication and information. Then let's explore the terms communication and information within a traditional business environment and then compare them to a total quality organization. Through such an approach, the differences in and importance of each should become clear.

Why are the subjects of communication, information, and training important to a safety and health program? Why are we talking about them in the first place? Quite simply, any effective safety and health program, whether or not it is total quality, will *live or die* based on these three components! In a total quality safety and health program that encourages change and improvement, communication, information, and training are even more important than that! That's why they are included.

TRAINING

So many critically important parts make total quality go in a program or an organization. Few, however, are more important than training. In a total quality environment, all levels within the organization must do more, carry additional responsibilities, communicate better, etc. You simply cannot expect anyone to take on additional responsibilities or interface differently without first providing the necessary skills for doing so. Traditionally, training has seldom been a forte in American business. Sure, most of the major industries and companies have had training departments, but training was only focused on key elements of a worker's job and dealing with labor-management issues. Basic skill building, people skills, communication skills, or skills in other areas such as safety and health were not addressed. They weren't considered important.

In total quality, the concept of training is very different. It is no longer a reductionistic concept, focused just on the worker or on worker-supervisor relations. In total quality, training focuses on the holistic realm, the working environment. It is both a microcosm—and a macrocosm—perspective. Training must meet the needs of total quality's dynamic process. Everyone within the working team sees new responsibilities and expectations in total quality. These require an expanding level of skills in work, communication, team dynamics, problem solving, presentations, etc. Training in a total quality organization must make the same quantum leap that the expectations for workers and line management do.

Total quality training must be diverse. It must be skill level dependent. What are the expectations for performance? The common thought that members of line management require a higher level of training does not hold true in total quality. Sure, a member of management must have people skills above that of a production or maintenance worker. The worker's skill level and knowledge of the processes or maintenance, however, exceeds that required by management. Because of this, training must also be target-audience oriented. Group training is fine in some areas, poor in others. It is dependent upon the skill level required.

Something is obvious here about the timing of training. Skills must be developed ahead of the expected performance, or at least concurrently with it. This requires that training be part of the planning process, along with changes within the organization, work team, or program. Training must keep ahead of the other aspects of the plan. It's a cause-and-effect concept. Training cannot lag.

One of these critical training areas is safety and health responsibilities. In a total quality organization, workers and line management have greater responsibilities for performing safety- and health-type duties. These usually include observation and correction of hazards or unsafe actions of workers, safety inspections and audits, and accident and incident investigations. These are skills learned via training.

Training within a total quality organization must meet both the knowledge and skills needed as new or transferred workers enter a working environment. This is *not* traditional new employee orientation where the new employee gets half a day from Human Resources, a safety talk and then gets turned over to his or her supervisor. Would you acclimate a new automobile driver to the Los Angeles freeway system in the same manner? No! Then why have we for so many years felt satisfied when we gave this amount of training to new employees? In a total quality environment, the speed is faster, more is expected, and greater autonomy exists. Therefore, you must have a much more extensive pre-placement training regimen, as well as a controlled, buddy-system introduction into the workplace. Also, don't forget the training requirements for the transferred worker, the worker who is entering a different work environment and team. Training is critical here and slow integration into the working environment is necessary.

More than ever, American industry relies on temporary or casual employees. Trends show that this move to contract employees will grow substantially in the future. Historically, these employees have been seen as the "disposable employee" from the "here today, gone tomorrow" vantage. Too often we have also viewed them the same way when it came to training, knowledge of hazards and precautions, and injury accountability. This isn't true in a total quality organization. It can't be! To do so would be too disruptive to the "everyone participates and is responsible" premise of a total quality organization. Any training program needs to focus also on this growing percentage of the workforce. Obviously, if extensive training becomes the rule rather than the exception, there is a point where temporary employees are not economically beneficial in relation to the investment involved in getting them up to speed.

Let's return to our analogy of the Los Angeles freeway system. With the limited amount of training we do with drivers, it is little surprise that we kill so many on our highways each year. We mandate safer cars, but shy away from building safer drivers. Think about it. How much retraining do we require of automobile drivers in America? Right! The same amount that we usually give our employees in our traditional way of doing business. On the freeways as in our factories, we train people once and expect experience to reinforce good driving and work skills. Does this work? From your own observations while you drive to work or home or from your own driving practices, how many drivers signal before they turn, change lanes, or enter or exit a freeway? How many drive faster than the posted speed limit? How many drive too slowly in the left lanes of the freeway? How many make unsafe lane changes? Why do drivers slow down when they see a police officer? The evidence speaks for itself. The fact is, traditionally, we have planned little for the retraining of our workforce. And, historically, we are surprised when they mess up. Consider that the rate of change is increased in a total quality organization. The obvious point is that retraining is not only important, it is critical in a total quality environment. Training *must* be regular!

Training:

- ▸ Must focus on the total work environment
- ▸ Must be diverse and skill-level dependent
- ▸ Must keep ahead of expectations
- ▸ Must include safety and health skills
- ▸ Must include new and transferred workers
- ▸ Must include temporary and casual employees
- ▸ Includes regular retraining

Training in a total quality organization is more fun. Because of the changes in the workplace, the training must be more frequent, imaginative, creative, and stay on the "cutting edge." There is little need for training-as-usual-type programs. The training conducted last year is often antiquated when complete. As the workplace changes and the skills of the worker increase, the training must meet those expanding needs. Training in a total quality organization is never ho-hum. It's exciting and challenging.

COMMUNICATION AND INFORMATION

American industry has not had the best record of communication and information. Rather, communication and information are filtered (or censored), one-way, adversarial-oriented, and viewed by management as high risk. Traditional thoughts concerning communication and information can be categorized into six areas. First, "Tell workers only what they need to know." This is based on a hierarchal-knowledge concept. The higher in the organization you were, the more you knew about what was going on. Workers at the bottom of the pyramid didn't need to know much. Second, "Some information workers can't understand." This dates back to a time when there was a large disparity in educational levels. Today, however, it is obsolete. Third, "Don't tell them the whole story." This information filter was supposedly for supporting business decisions or to simplify complex issues. It didn't work then. And, with all the attention today about business ethics and the complexity of life, this thought doesn't hold up. Fourth, "Sugar coating is always best." Management did what your mother did—provided a filtered, rose-colored view. It's no wonder that so much suspicion exists between labor and management today. Fifth, "Avoid open forums." This was both a "bob and weave" technique and one that avoided common knowledge and worker synergy. With the critical need for everyone in your organization to know the same information, avoiding open forums should be less common today. And sixth, "Communication is best on management's turf." It's a classic "my ball" and "my yard" security issue carried over from youth and adapted to the industrial environment.

Do any of these common traditional thoughts concerning communication and information sound familiar? Unfortunately, these philosophies have great followings in today's business. Those that embrace the traditional "business as usual" attitudes still have these thoughts buried deep within their culture. As an organization moves toward total quality, however, these thoughts are replaced. Even in the most advanced total quality organization, however, one will find that some "Old World" attitudes hold sway somewhere in the organization. The industrial culture of a country takes a long time to change completely. Until all business moves into total quality, it is unlikely that this culture will totally change either.

What are the results of this traditional way of thinking about communication and information? There are three major ones. First, this attitude drives communication underground. Communication, instead of being overt, becomes covert. The rumor mill and the "grapevine" become *the* methods of communication. Unfortunately, those are not always accurate. Therefore, management wastes time chasing shadows or correcting misconceptions after the fact, in a highly suspicious environment. Not very productive. Second, this way of thinking creates workforce polarization. And this polarization is highly dependent upon the way we communicate *to* (not with) workers, and the way management views workers' need and ability to know and understand. If these barriers are high in an organization, polarization between management and labor is great. If these barriers are lower, the polarization is less. The third result is that it retards or poisons team and individual participation. Knowing that participation is key to a total quality organization, this result alone will poison any attempt to implement total quality.

THE TOTAL QUALITY RULES OF COMMUNICATION

How do these traditional thoughts concerning communication and information differ in a total quality organization? I've included ten rules for our discussion:

Total Quality Rules of Communication:

1. Communicate as much as you can
2. Communicate often
3. Communicate honestly
4. Communicate quickly
5. Communicate in many forms
6. Communicate on their turf
7. Invite open forums and questions
8. Open doors are the only kind
9. Too much information is far better than too little
10. Communication is a two-way concept

Communicate as much as you can. I've become painfully aware that communication is *not* a natural ability. In reality, we avoid communication. We build fences around our houses. Look at the high divorce rate! We greet others with statements such as "How are you?", but we really don't want to know. When we get into situations where we need to really communicate, we clam-up and feel very uncomfortable. When employees want to know what's going on, we either avoid them or give them a superficial answer. Communication isn't a natural ability. So, when I say that the number one total quality rule is to communicate as much as possible, realize that this is easily said, *not* easily done. There is a strange confidence-building, barrier-destroying charisma that surrounds those who can communicate. Jack Kennedy had it. He spoke and we all listened. Whatever he said, we, as a country were willing to try. Martin Luther King, Jr., had it too. That's the magic of communication. That's also why it is so critical in a total quality organization.

It's a simple equation. Communication builds trust. Trust leads to commitment. Commitment results in participation. You *cannot* build participation without trust, nor can you create trust without communication. There is another truth about communication. Like feedback, the amount of communication that is needed can only be determined by those who need the communication. It isn't the communicator but the receiver who determines when "enough's enough." That's why communicating as much as you, the communicator, can is critical to the total quality process. Communication doesn't focus on meeting *your* needs. It focuses on meeting *their* needs.

Communicate often. "The silence was deafening." Ever heard that saying? It comes when someone is expecting to hear something and doesn't. It would be nice if we all had a Communication-Needed light on the top of our heads. Then we, as communicators, wouldn't have to guess when we need to communicate. We would communicate when the light came on and stop when it went off. However, we don't have a Communication-Needed light. To combat this, we need to communicate often. Remember—we don't communicate to meet *our* need. We communicate to meet *their* needs.

Communicate honestly. Have you heard the saying, "One Ah-Shit wipes out 1000 Atta-Boys." It's a slang expression that says that one mess-up will remove all your victories in one fell swoop. Gone, right now! There is no area where this is more true than in communication. Say one non-truth, *or* have what you say perceived as half-truthful or untrue, you become a liar and any future communication is ignored or not believed. The honesty of your communication determines what you're worth, your character and ethics. It's true! For example, what is your perception of politicians and used car salesmen?

Communicate quickly. This causes much heartburn in traditional upper management. There is, however, a pervasive monster in every workplace that cannot be ignored. We call it the "Rumor Mill" or "Grapevine." Rumor mills are extremely fast, and once a rumor is on the vine it is difficult to change. That's why communicating quickly is so important. If something changes or something occurs that needs communicating, do it quickly and avoid the "Grapevine Blues." Don't wait to communicate until you've thought about it, get a consensus decision, or get a feel for "the water." By then, it's too late! An organization needs to develop a certain level of comfort for, and not fear of, immediate communication. The first time someone "loses his or her head" because he or she spoke before being authorized or approved, the "Grapevine" will shake and cause more ripples to your total quality process than you could imagine.

Communicate in many forms. Don't assume that using one form of communication will be effective and timely. Bad assumption! One of the forms of communication that management uses the most is the written-type, the infamous memo. In reality, written communication is a very poor form. Why? It's slow. It assumes that everyone can and will read it. It assumes distribution to all who need to know. It is often misinterpreted. It is one-directional. And, it very often gets lost, set aside, or misplaced.

Verbal communication is probably the best form of communication but also has drawbacks. It is limited by the number of people you meet face-to-face. It can be different from group to group. It is easier to make mistakes in verbal communication than in written communication. So, when I say to communicate in many forms, the purpose is to compensate for the drawbacks inherent in different forms of communication as well as the assumptions we place on them.

Communicate on their turf. Why do we hide in our offices as much as we do? It's because that is where we are most comfortable and protected. This is the same feeling workers have. Their turf is where they are most comfortable, not in your office or in a meeting room or auditorium. Their turf is where their working support-group exists. Their turf is where they are most attentive and where the greatest opportunity for two-way communication exists.

Invite open forums and questions. This a frightening one, especially if the subject is controversial, negative, bad news, or focuses on change. When considering small groups or open forums, the question is this: Would you rather face the lions once and make sure that everyone hears the same information, or would you prefer to battle the monster over the next week, month or year, and possibly never get beyond it? Sure, it's frightening! Looking at the alternatives, however, it's necessary.

Open doors are the only kind. Does your organization have an "Open Door Policy"? Is it a reality or a myth? Do your workers think that if an open-door policy is used, "Bad News" will be back to their supervisor before they return? Open-door policies are great, but they must be real, confidential, honest, sincere, and known for positive results. Otherwise, the "open door" is a fiction.

Too much information is far better than too little. You want people to be able to throw information away because it isn't needed, not create lines because the needed information is not there. It's the truth. There is no limit to imagination and creativity in questioning minds when they are left to "fill in the blanks."

Communication is a two-way concept. Too often, especially in upper management, we fall victim to the thought, "I talk, you listen." Kind of the way we talk *to* our kids. Total quality invites an open dialogue. It *has* to be a two-way street. "I talk, you listen. You talk, I listen."

IT MUST BE DONE WELL

In a total quality program, training, communication and information cannot be forgotten, and must be done well. To work, total quality depends on a high level of employee participation and involvement. Participation and involvement are impossible without, first, giving your workers at all levels the skills, knowledge, and information they need. Realistically, it shouldn't take total quality to tell us this. It should be common sense and automatic. Unfortunately, if you look at how we have managed our businesses in the past and today, it isn't. This is a product of our upbringing and our culture. Whatever the cause, in total quality organizations, all three must become strengths throughout the organization, from the top to the bottom and across all functions. Safety and health is not isolated or set aside from this fact. We must be active in identifying the necessary safety and health training. We must be continually aware of the importance of communication and information. And, we must be active participants in the communication process.

Want a simple, foolproof test of whether a program or organization has embraced total quality or is its quest just "lip service"? Look at an organization's or program's training, communication, and information network. What is the mission and goal of the training program? Do they communicate quickly and openly or do they ignore the need? Is information a priority or something they feel they "have to do"? Ask them two questions: "What worker training are you or your organization doing right now?" And, "How do you communicate with your employees?" The answers will tell you whether or not they have embraced total quality.

13

THE MULTIFUNCTIONAL PROFESSIONAL

Around 1900, Frederick Winslow Taylor,[9] defined a profession that was previously unknown. He studied workers like those who shovelled coal in the Bethlehem, Pennsylvania, iron and steel industries. What he discovered was amazing to those in industry. By studying those who were more productive and efficient than others, he found that through minor tool changes and training, he could increase productivity. Needless to say, his work became very popular. Many tried to copy it, and with it, enjoy his success. Because he used some specialized techniques and was well skilled in observation techniques, however, it was difficult to repeat his success. What Taylor did was to pioneer a new field—the field of industrial engineering. Because it was so successful at increasing productivity, the profession boomed.

If we study that profession's history, we see that a strange thing happened. Because of the demand and the techniques used, the field of industrial engineering became more and more specialized. This specialization required a high level of training and skill. As it became more specialized, it moved from the factory floor to offices. Thus, it became a staff function. This was ultimately destructive to this profession. Why? Removed to offices, the industrial engineers lost touch with what was really happening in the factory. The profession became so specialized that its focus changed. It sought to advance the profession and maintain

---

[9]M. Greif, *The Visual Factory: Building Participation Through Shared Information,* (Cambridge, Mass.: Productivity Press, 1989).

the turf it had gained, not to improve industrial productivity as it was originally intended to do. Instead of looking outward, the profession began to look inward and "delight in itself." The profession became more important than the original mission.

What happened as a result? For the most part, industry became disenchanted with the field. The number of practicing industrial engineers became fewer and fewer. The demand continued to decrease. Look at the change in the number of practicing industrial engineers as a function of the total number of people employed in business today. The drastic decrease makes this truth painfully real.

Today, as a result of both Japanese manufacturing concepts and total quality, industrial engineering is rebounding. The resurgence, however, is not picking up the profession where it left off. Instead it focuses on Taylor's original work. The skills of industrial engineering are being applied on the factory floor, within the line structure, not in the Office of the Industrial Engineer. These same skills Taylor developed are taught to managers, supervisors, and workers alike, and they are empowered to use those skills for continual improvement. Through these empowered and skilled workers, great advances in productivity are made, quickly, inexpensively, and easily with minimal resistance and roadblocks. Has this new approach to industrial engineering been successful? You bet it has!

WHAT CAN WE LEARN FROM HISTORY?

What can we, as safety and health professionals, learn from Taylor and the profession of industrial engineering? Has the anvil dropped on you yet? Do you see any parallels here?

The practice of industrial safety didn't begin as a highly educated and skilled, staff-oriented profession. It began on the factory floors across America. Principles of accident prevention were developed and implemented by foremen, supervisors, and workers alike. It wasn't the Safety and Health Professional. It was the "safety guy." It was a grass roots concept. As the profession became more and more complex, the profession moved from the factory floor into offices. You know, the

"Safety Office." And, like industrial engineering, it began to lose touch with the industrial world and more important, its purpose.

That's a bold statement. Throughout our history, we have used many words that sound like this shift hasn't happen. However, I challenge you and your Safety and Health Profession Paradigm. Consider this question. Today, is the safety and health profession more interested in itself or improving worker safety and health?

Consider these arguments. How many members does your professional association or society have? What are their membership growth objectives? Do these membership pushes focus on increasing the "clout" of the profession or improving worker safety and health? Consider the argument about the "competency" of certified versus noncertified safety professionals or industrial hygienists? How about the arguments that focus on licensing professionals? Are certifications and licensing aimed at improving worker safety and health or at keeping the profession pure by regulating or self-policing? What's the growing portion at the professional conferences: "how to" subjects for new or learning practitioners or other subjects like licensing, ethics, affecting governmental actions, etc., for professionals? Is the person just entering the safety or health profession granted full membership in our professional societies or associations or does he or she have to "qualify" by having a degree or extensive, documentable experience? As professionals, who do we admire the most, the obscure, basically educated safety or health practitioner who spends his or her entire career on the factory floor helping a limited number of people or the highly educated and credentialed corporate director of safety and health, the supposed leader in the field? Have our professional certifications become so important that they are inseparable from our name? Do they appear on our office nameplate, on our business cards, or on our name tags when we are around mere practitioners of the profession?

I hope I haven't irritated you. That isn't my purpose. Have I hit a spot close enough to your Safety and Health Profession Paradigm, however, to make you think? Hitting close to your paradigm is important because it is necessary for you to get emotionally involved to understand this important point. Like the profession of industrial engineering, we are in danger of "delighting in ourselves."

In my opinion, and in the opinion of an expanding percentage of the safety and health profession, this has become one of the major barriers to advancing the original safety and health mission—improving worker safety and health. Instead, it has become the major factor or power for the protection of the profession. Today, it is the major force that brings us to the insidious and incorrect belief, the fear, that it is "them or us," "safety *against* quality." When, in reality, this fear only exists in our paradigm. It isn't "them or us." It is only "us," *all* of "us." We must learn from history, particularly the profession of industrial engineering. We need to wake up and change our paradigm, because, like industrial engineering, *we* are in the position to lose the most. Think about it! Think about the supportive evidence you see around you. In a RIF, who is more apt to be "out-placed," the highly educated and specialized safety and health professional or the quality person?

TOTAL QUALITY INTEGRATES
SKILLS AND KNOWLEDGE

Let's focus on one of the major, underlying concepts of total quality. Total quality isn't about protectionism of *any* group or function. It's about the blending of skills and knowledge to form a greater synergy and achieve greater results. Those who cry, "We need to become as important as the quality folks," or, "Substitute the word quality for safety," don't get it yet. In total quality, and for the betterment of our mission for worker safety and health, it isn't a competitive, win-lose picture at all. It isn't about preservation of a profession. It's about survival of a greater cause, our mission!

We've become engulfed in this belief. We have tried for so long to "get some respect," "keep the profession pure," or "make the profession grow," that we have not looked to the edge of our Safety and Health Profession Paradigm to see the future. The future isn't in the protection of the profession. The profession isn't what's important! Our mission as safety and health practitioners, the safety and health of our workers *is*!

PROFESSIONAL PURITY

Challenging our paradigm causes us to question one of the most basic beliefs in our profession. This belief says that learning new disciplines or taking on new responsibilities will dilute the safety and health effort. Wrong! It only dilutes our concept of our profession's purity. In fact, it has the potential to enhance our mission! You see, by diluting our "professional purity" and taking on new responsibilities in a total quality environment, we come to a choice. We may choose to do less safety and health. You know, "Well, if they want to dilute my job in safety and health, they'll see what the costs really are." It's the classic "cut off your nose to spite your face." The second choice available to us is to embrace the change and move toward the participative side of total quality. There are definite advantages to this. By spreading the responsibilities for safety and health back into the line structure, back onto the factory floor, we build true participation. At that point, it's a win-win picture. You see, it isn't "us or them." In total quality it is only "us."

It may take a little time to contemplate a thought such as this. After all, it is opposite the traditional way of thinking. If you move on too quickly to the discussion of multifunctionality, it will only add more confusion because, confused, the mind absorbs little. As you reach this point, you may choose to take a break. Maybe a moment of thought is enough. Maybe you might need an hour or longer. Before you go on, however, let what I've written percolate around in your gray matter for a while. Then we can look at why this has happened and what opportunities lie ahead.

THE SEARCH FOR RESPECTABILITY

The safety and health movement, continually sought respectability throughout its history. Like Rodney Dangerfield, we have been seeking a little respect. More than respectability, however, it has been an up-hill battle to gain equality for safety and health in the workplace. For years we have seen slogans like, "Safety is Number 1" but never felt like it was.

Time after time decisions between the safe way of doing something and getting the product out the door were most often answered by, "Do it as safe as you can, but get the product out."

Philosophically, it was always arguable that, if nothing we do is safe (absent from harm or risk), safety is merely a matter of degree, determined within each instance. Productivity is easy to quantify. One widget manufactured and shipped always equals one widget on the order book. Quality is also more easily quantified because it is defined by process or product dimensions or specifics. Safety and health isn't easily quantified. Sure, everyone wants to be safe, but without a way to quantify it, it always seems to be third in importance, not first.

Historically, the responsibility for safety and health has been placed on the "safety guy" or department. Because of this, our challenge has been to bring about this equality between production, quality, and safety and health. This is where we lost our focus. We began to focus on the importance of the "Safety Office" as the banner carrier for safety and health, not on the environment that created the inequality. It became a power struggle, not an equality issue at all. To help us understand, let's look at that environment and the complications it causes.

The job of safety and health is normally placed (or buried) within the staff structure of an organization along with other competing staff functions such as employee relations, environmental, accounting, marketing, engineering, quality assurance, purchasing, etc. With the flat structure, all staff functions find themselves in an inescapable "dog-eat-dog" environment. Safety isn't alone in trying to find equality through power. Each staff function has to continually vie for respectability and the ear of upper management. It's a "turf war." This traditional "it's them or us" way of thinking places enormous stress on all involved. Each wants to gain favor and climb to the top of the heap. Because we are all stuck in this reality, it's no wonder that some ways of thinking—paradigms—have become deeply imbedded in our thought process. One of these has been, what I call the "Purity of Practice Paradigm." The thought is that by elevating the profession outside of the workplace, we gain power at work. This has never worked because we focused on the practice or profession as a power base, not on the cause of the inequality in the workplace.

Look at a simple analogy of this error. This approach is like trying to fix a quality problem in your plant's production of widgets by trying to increase the importance of quality in America. Maybe we can pass some quality laws and create quality police. You know, like the OSHA Act and OSHA inspectors. Would this approach to quality have a direct and sustained impact on the quality of our widgets? We'd go out of business because we lost customers due to poor quality long before we'd know!

SUCCESS *VERSUS* POWER

Another supportive paradigm is what I call the "Cookie Cutter Safety Paradigm." I call it the "Cookie Cutter Safety Paradigm" because of its supposed application to all workplaces, all organizations, all cultures. It's a "one size fits all" paradigm, like using a cookie cutter to make hundreds of identical cookies or safety and health programs. Within this paradigm, we have listed the components of a successful safety and health program for many years. We've come to believe that these components of a successful program are akin to the Ten Commandments, chiseled in stone. They are flawed, however, because this paradigm runs the same path that the Purity of Practice Paradigm does. It is focused at creating a strong Safety Department, a power position.

What is commonly conveyed as the number one element of a successful safety and health program? The number one aspired-to element has always been support from the top. From a total quality perspective, there are two interpretations of the word "support." The first is support of the safety and health office and its policies or rules. The second is support for worker safety and health. These are *not* the same. Based on both of these common paradigms, it appears that we have inferred the first meaning. After all, the first meaning reinforces both paradigms. This support is the power behind the Safety Office or Safety Director. It is responsible for the "verbatim compliance" attitudes concerning safety policies or rules. Does a strong safety and health department equal a strong commitment to worker safety and health and, consequently, a good safety program? Unfortunately, it does not. From the traditional approach to safety and health, the Purity of Practice Paradigm, and the Cookie Cutter Paradigm, however, this correlation between a strong safety department and good worker safety and health is still believed to be true *and* practiced.

This belief has led traditional safety and health departments to confuse importance of the mission with turf. It's common for safety's importance in an organization to be considered proportional to the size of the safety staff or the size of the job. The bigger the safety and health staff, the more important safety was in the organization. How have we responded to this belief? Given a new responsibility, expanded duties, or new regulations, the safety and health department sent in a requisition for another person. The message to upper management was "more work demand = increased turf." Vying for the same turf and respect, it's no wonder that our relationships between other staff functions are strained. It is also no wonder why upper management turns to the safety and health group when a RIF comes along. Memories are short, particularly in a financial crisis. Safety and health is usually fat. Not perceived as good team players, no one else really cares if employees are cut out of safety and health.

The traditional approach to business shows a clear delineation between line and staff safety responsibilities. It goes like this: The line organization issues discipline and the safety and health group issues rules and inspects for compliance. Within the staff-oriented safety and health department are the highly trained and educated experts. Because they are experts, they approve drawings, procedural changes, new rules, interpretations, etc. Everyone has to go to them. In a power position, safety people look at line management as beneath them. Line supervision looks at safety people as elitists and roadblocks. Therefore, animosity between safety and the other staff functions is compounded by friction between safety and the line structure.

The power of the safety and health function is measured by the size of the job. More responsibilities equals more power. These safety and health responsibilities have included performing inspections for compliance, providing safety training including the infamous safety orientation for new employees, performing the accident investigations, and establishing safety rules. These responsibilities of the safety and health function are pretty standard.

This responsibility = power belief has had some interesting side effects. First, we've created two different and conflicting "piling-on" principles, one internal and one external. Internally, new responsibilities are invited by safety and health because more responsibilities equals more power and, possibly, more staff. The responsibilities have to be "safety- or health-related" to be "power-based" though. That is the internal catch. Remember the Purity of Profession Paradigm? Externally, however, adding responsibilities outside of safety and health is met with resistance. Why? We don't want to dilute the safety and health effort. Strange paradox!

The second effect is that because we have become very specialized, we've become less visible to management and to the people on the floor. Because of this, a perception that no visibility equals no value to the organization results. This forms additional ammunition for targeting safety and health staff in RIF situations. Remember industrial engineering?

And third, there is a tremendous challenge to keep turf. Winning today will not ensure winning tomorrow. The competition for power never ends.

What are the results of these paradigms and beliefs? The quest for power has become *the* battle. Fighting for it has become a totally focused activity. Safety and health is always vying for support and measuring it in terms of turf. It is commonly known as politics. In short, the practice of safety and health is stymied within its own paradigms and misplaced focus. It is stuck in the 1950s instead of moving into the 2000s. These paradigms and our continual allegiance to them are resulting in the profession moving forward, not the safety and health mission. We've become mired in the profession. If we are to move forward in our mission as safety and health practitioners, however, something has to change.

A NEW PARADIGM IN TOTAL QUALITY

Total quality brings a new way of thinking, a new paradigm. That paradigm says there is no longer a reason to vie for equality or power.

Some of us will continue to think as we have for years. Some fewer of us will change our paradigm. Those who change recognize that the paradigm has shifted, as has our world. Those who change their thinking will be free, free from fighting the shadows cast by the past and free from battling the equality issue, which only existed because of the way we were thinking. Thus, we will be freed by our new paradigm and our new way of thinking.

How is this new paradigm different? For us, the safety and health practitioners, the new paradigm changes our focus from responsibility to coordination. Under the new paradigm we no longer focus on the frustrating staff-oriented responsibilities of safety and health, only the coordination of an organizational safety and health effort. That, in itself, removes a significant power-seeking burden. This new role removes an even heavier burden from us, one that we have carried and misapplied for years. That burden is having responsibility without direct authority. The new paradigm requires new skills, though. Skills that we must learn and in which we must become proficient. Only this way can we be champions of our mission for worker safety and health.

Line responsibility for worker safety and health is fundamental to total quality. This is where it should have started during the industrial revolution, but didn't. This safety and health responsibility lies with *both* management and labor. Most traditional Safety and Health Office responsibilities transfer to this line responsibility. It creates an environment where responsibility exists with direct authority and where principles are emphasized, not rules. Safety and health is allowed to reach new heights through line empowerment and participation. All employees within the organization, including the safety and health practitioners become part of the total quality turf and let go of their own.

What are the perceived risks of letting go of the safety and health turf? These risks are a reason that some safety and health professionals will not be able to make this transition to the new paradigm. Measured through the old paradigm, the perceived risks include the apparent loss of importance. It's an ego issue. Another risk is having to work with and through others instead of doing it ourselves. Historically, safety has never been considered a team player. We have insisted on having only our way or perspective seen. Total quality is a real change. A trust issue can be seen as a risk. In the new paradigm safety oversight is almost totally eliminated. Some perceive this as an unacceptable risk. Again, it's a trust issue. Another risk is that it brings a lot of "sacred cows" up for discussion and, perhaps, change. It also calls for sharing the information *and* knowledge about safety and health. For some, the thought of someone else, especially in the line structure, making safety decisions is extremely frightening. The new paradigm calls for letting go of some, if not most, of our acquired powers. Team play is very frightening because we lose direct control. In total quality though, we place control for worker safety where it should be, in the line structure.

What happens if we can make this change in paradigms? One of the most notable is that we suddenly have time available. The phone rings less. We have fewer "fire drills" to attend to. We have the opportunity to change from reactive to proactive safety. We can then plan what we do. These are not bad trade-offs.

The new paradigm allows us to expand our repertoire by learning other skills. This, in turn, reverses the importance versus turf challenge. We need fewer safety and health staff, not more. Overhead, thereby, is reduced, and we become heroes. And, we experience better coordination of safety and health through expanded knowledge and fewer turf conflicts.

The "boom or bust" world caused by enforcement or new regulations becomes extinct. There is no more need to base size of the safety and health staff on the number of regulations. The total quality approach "level loads" the staff in line with Just-in-Time (JIT) principles and transfers the burden to training, coordinating the effort, having the responsibilities accepted within the line structure, and using control programs. This is a very different world from what we have traditionally known!

CONFLICT OR PARTNERSHIP

What about taking on new responsibilities such as those involved in other regulatory areas? Do these dilute the professional's effectiveness or expand his or her horizons? Good question. If you poll any group of safety and health professionals, you will find equally immovable camps on both sides of this issue. One camp feels that taking on other regulatory responsibilities gives them more, and more thorough, knowledge. They feel that it's a real advantage. The other camp, the larger group, feels the opposite way. This group is always much more passionate and protective of its feelings about "diluting efforts."

Have you ever stopped to think why taking on the responsibility in other regulatory areas meets with such inner conflict? I have. It comes from the familiar, traditional safety and health environment. Because we think "your job" versus "my job," we're not able to escape the triggers that cause this feeling. From upper management's perspective, it goes like this. "The safety and health guy isn't busy." Loosely translated this means, "He or she is not visible to me. Isn't keeping workers safe and healthy an easy task? Aren't all regulatory efforts the same? If it's a regulatory issue like transportation, ask the safety guy." The safety and health person sees it differently. We are always too busy trying to catch up, put out fires, and get the important jobs done via our normal reactive management style. From this perspective, it goes like this. "I'm too busy now. I have no more time to do anything else, much less learn more regulations. Besides, I'm frustrated. I have no interest. I'm only one person." It's no wonder that merely mentioning additional responsibilities to a safety and health practitioners in a traditional environment and way of thinking brings on an overwhelmed and overworked feeling.

In a total quality organization those responsibilities are more easily merged. The safety and health person in a total quality organization is not less busy than those in other environments. He or she is, however, busy in different ways, proactive versus reactive ways, coordinating rather than doing. Also, to a total quality safety and health practitioner, perspective and knowledge are everything! That's not to say that additional responsibilities will not require balanced and additional efforts, but they also will reduce wasted efforts and avoid conflicts, too. Let's see why.

First, let's look at the dark side. There is, of course, a negative side to viewing both sides of any regulatory issue. Before, it was always a power struggle, the safety and health guy against the environmental or transportation person. This creates several different possible scenarios. For example, a chemical used in the workplace like 1,1,1-trichloroethane, can be good for safety, but bad for the environment. It is nonflammable, evaporates quickly, and has a relatively high exposure limit. On the other hand, 1,1,1-trichloroethane is a listed hazardous waste, is land-banned by the EPA and requires expensive incineration for disposal. When it gets into anything else, it makes it hazardous too. The second possible scenario is to use a chemical that is less hazardous for the environment but presents a significant safety risk. A good example is substituting ethanol or ethyl alcohol for 1,1,1-trichloroethane. Ethanol is a natural product of plant decomposition from anaerobic respiration called fermentation. It evaporates rather quickly and is much less harmful. Ethanol is natural and presents a minor environmental risk. Safety wise, however, it is highly flammable. If you use ethanol in greater volumes, it poses an unacceptable safety risk. A third scenario could be that you use a material or chemical substance that is not harmful to the environment and is OK from a safety perspective, but falls into a regulatory maze. You can't ship it, transport it, or treat it. Certain reactive chemicals fall into this group. How about a fourth scenario? A waste is bad from both environmental and safety perspectives, but you can't get rid of it. An example is mixed wastes that contain both hazardous or toxic chemicals, and are radioactive besides.

These problems come from the discrepancies that lie within the regulations and different ways of assigning risk. Each regulatory agency has its own private agenda and perspective. That is complicated because the regulatory agencies don't communicate among themselves. Why should they? That would only reduce *your* conflict. Regulatory agencies are also influenced by special interest groups. Thus, some of their regulations are more fear-motivated than fact- and research-based.

If these regulatory responsibilities are separate within an organization, additional areas of conflict arise. Even the best organization has communication difficulties. Usually, these are the result of different reporting structures, different priorities or projects, not knowing what or what not to communicate, personality incompatibilities, or politics. An elevated form of these communication difficulties is called "turf conflict," where functions actually *choose* not to communicate or cooperate with each other.

Tunnel vision exists within any group. Most of the time it's unintentional. It comes mostly from our concentration on a task or project.

We get so tied up in it, we forget to communicate. There can also be conflicting policies and solutions. This can be messy where the safety policy says no and the environmental policy says yes, or vice versa. Negotiating a settlement is not possible until the policies are changed. And there is always the potential for one to be insidiously worked against the other. This is usually an action arranged by someone outside both groups seeking to play both ends against the middle to ultimately get what they want. Or it can be an unknowing act where there are conflicting priorities, a common project, and no communication.

IT JUST MAKES SENSE

Total quality works to eliminate wastes including those brought about by communication, turf, policies, or priorities. In a total quality organization, combining the regulatory efforts under one person or group just makes good sense. Why? It has some definite advantages. For example, within a single group (or professional) communication is much better. There is also better coordination of resources. Coming out of one office, the potential for conflicts in policies and solutions is much less of a problem. An important advantage is group or team resolution of conflicts and differences of opinion. This avoids public conflicts or the all-too-famous memo wars. Having more consensus also saves costs by minimizing wastes, including time. There is just better integration and coordination into the total quality processes.

Merging regulatory coordination makes good sense from both a total quality perspective and a line responsibility perspective. It simply causes less confusion when the direct responsibility for safety- and health-type activities and environmental compliance fall within the line structure. It avoids the "tennis match" neck cycles where line management turns to first listen to the safety perspective and then turns in the opposite direction to hear the environmental position. One voice to line management is always the preferred way.

TEAM SAFETY AND HEALTH

Please don't confuse this topic, team, with the all-too-familiar concept, committee. "Safety by committee" has been the flagship of safety for a long, long time. It's the classical approach used by upper management. When you want to improve safety, management forms a Safety Committee. These committees always have common compositions and problems.

Historically, we use three different designs for Safety Committees. In some organizations, the Safety Committee is made up of management only. The rationale is that only management can correct safety and health problems. The second common makeup for a Safety Committee is workers and the safety person. This is a "grass roots" concept that thinks that talking about safety problems and concerns at the worker level will provide a valuable release and camaraderie. In response, it is thought, the workers will begin to work more safely. The third committee composition, 99 percent of the time, is driven by a labor contract. It is the Joint Union and Management Safety Committee. Elected or appointed members from labor and members from management regularly sit around a table and discuss "safety problems."

Have any of these Safety Committee structures worked? Unfortunately, they have all struggled. The management-only committee suffers because it loses floor-level perspective. Management can only guess what the safety problems are and what will best solve them. This causes a lot of wasted energy and money working on perceived problems that aren't really problems. Or implementing solutions that either won't work or aren't accepted by the workers. So, management-only Safety Committees are out of touch and that makes them ineffective and wasteful.

"Grass roots" Safety Committees haven't fared much better. Why? Mainly because they lack the critical communication, input, and involvement from management. Significant restrictions are always placed on worker Safety Committee authority, too. It is also flawed by the misplaced expectations of management that it will be a release to employees and make workers work more safely.

Neither of these one-sided committee structures works because they are one-sided. Safety has to be a two-sided effort. Labor realized that years ago. To solve it, contractually, they pushed for and gained joint union and management committees. Have they worked any better? Usually not. Why? Because, they are divided by the very instrument that creates them, the labor contract. There are always "your issues" and "my issues," "your way of looking at it" and "my way of looking at it," "your responsibilities" and "my responsibilities," "your loyalties" and "my loyalties." They aren't "us" oriented. They are locked-in the unwinnable situation of "us" versus "you."

How is team safety different? Think of the difference this way. Imagine a sunny Sunday morning and there's THE BIG football game with national television coverage. All the big-name football commentators are there. There is hoop-la and pre-game activities. It's exciting! Picture it in your mind. The announcer begins to introduce the game. "Welcome everyone to today's exciting game, the one you've been waiting all season for, the Safety Bowl. Today's big game is for the championship, pitting the Washington Redskins Football Committee against the powerful Seattle Seahawks Football Committee."

Football *what*? What the hell is a football committee? It's supposed to be a football TEAM!

Capture the thought that's running around in your head right now. What's wrong with your mental picture of a football committee? Because within that mental picture lies the reason committees don't work. Think about it. When you picture committees, you see people talking, perhaps arguing, but talk-talk-talk-talk, little *do*. Why else do you think that you resist when your spouse volunteers you to be on a school committee or your boss suggests that you should become part of a committee at work. "Committee? Yuck! Committees are all talk and no action. I've got better things to do with my time. Committees are just a waste of time! Committees meet for the sake of meeting!"

Remember when you were in elementary school or junior high (middle school) and they picked sports teams in the physical education classes. The coach would say, "OK kids, listen up! Today we are going to be playing softball. You and you are the captains. Flip a coin and see who picks first." One captain and then the other would pick the kids to be on his or her team. There was a lot of active discussion and buzz between those already picked and their captain about who should be picked next. Those not yet picked, however, were very quiet. Standing with the unpicked, remember that uneasy feeling, that tingle that went up and down your body, excited about the possibility that you would be next, or more so, worrying that you would be picked LAST! Why did you get that feeling? Because there is a power, a deep need that goes with belonging to a team, even for just a class period. Belonging to a team is *everything*!

Why is belonging to a team so much different from belonging to a committee? You've got to be kidding, right? Teams are very different from committees. Even your mental picture of a team is very different. You picture involvement, excitement, working together to win, cheering for each other, esprit de corps. A team conjures a vastly different picture from the one we get with the word committee, doesn't it?

The word team is in use more and more in our working world today. I have a question. As you have experienced teams in your organization, have they agreed more with your mental picture of a committee or of a sports team? Probably it more closely agrees with your mental picture of a committee, doesn't it? That's sad! Because what has happened is that we have only changed the words, substituted them. And in doing so, we have naively assumed that by changing the word, it changes the way people work. You have to change the culture before you can change from a committee-based to a team-based organization.

EVERYONE GETS INVOLVED

Total quality forces that change because it emphasizes participation and ownership. No, the bond with total quality is much stronger than that! Without active participation and ownership at all levels in an organization, total quality cannot exist! Participation and ownership are fundamental in team involvement. They aren't in committees. That's why teams work and committees don't. Teams are fundamental to total quality safety and health programs. The secret is simple. *Everyone* gets involved!

Workers must have ownership in safety and health. In a nonunionized organization, building and encouraging this ownership are easier. First, it's easier because there is no contract which impedes or slows this change. Second, there is no "middle man" through which change and communication is channelled. And third, there is no history that divides the organization into "us" versus "you" issues. In an unionized shop, it isn't impossible, it's just not as easy. There is a positive aspect, however. Once ingrained into a union shop, the new culture has greater staying-power.

There must also be line ownership in safety and health. Building this can be more difficult. It requires a culture change in the line structure that is actively and consistently championed by upper management. It has to come from a strongly held vision at the top. And the reality is that some members of line management may not be able to understand, change, and thus, survive in the new culture. That's a pretty strong statement but one that must be accepted by upper management for line ownership to occur.

For team participation to happen, there must be a high correlation with training. Team participation skills must be learned at *all* levels. Such an organization obviously places extreme importance on the effective use of control programs in safety and health. Tools that build participation are also important. Some participation tools are easy to use such as formalized programs you can purchase. The first one that comes to mind is the DuPont STOP™ program. The STOP™ program focuses on unsafe actions of workers and uses an easy workbook and video tape format to teach observation and hazard correction skills. It is a successful top-down program that has been around for a long time. There are other programs that have common aims—building participation in safety and health. There are also involvement tools you can use through management to get participation from employees. These include awareness building tools such as posters, video programs, pamphlets, booklets, etc. Workplace meetings on safety topics that are presented by workers or work teams are also very successful at building participation.

Historically, we have used safety awards as "carrots" for worker safety. Most times, these focus on workers staying injury free, not on worker safety participation. It's for this very reason that these injury-free-based award programs have mixed results. Participation-based awards, such as department awards for high safety participation, are different. They can significantly build employee participation. When used, they can change one of the perceptions destructive to safety award programs. That is, they bring a halt to the reward systems that depend on not having injuries and, instead, focus on involvement. Which is more positive, important, and longer lasting? Which is a direct result of the other?

The last tool you can use to build participation is probably the most effective. Tie individual safety accountability to each worker's compensation program, including management and workers. So those who are highly involved in safety and keep the participation high are rewarded with pay increases. Pay is the most effective "carrot."

Multifunctionality also builds participation. It is a key concept of Just-in-Time (JIT) manufacturing and total quality. At first, multi-functionality is seen as a negative. It doesn't have to be negative, and we certainly shouldn't fear it. It does compromise power bases. It requires a change in our paradigms and demands more communication and coordination skills. Multifunctionality works through worker participation and teams. It is the way of the future for *any* staff function, especially that in safety and health. It reduces waste and makes all feel more valuable to the organization and effort. It offers more rewards than it threatens risks.

In total quality, the real question isn't will we become multi-functional? Those decisions are made for us by upper management. Call it "right-sizing," "re-engineering" or any other term. It will happen! The real question is what can we learn to make the transition smoother and advance our mission? We need to embrace multifunctionality if we are to be successful in the future. Our traditional paradigms force us to look at it differently, negatively. Changing those paradigms, however, allows each of us to appreciate the real freedom, the new challenge, and the success that lies ahead in total quality.

14

SAFETY AND HEALTH
RULES AND ENFORCEMENT

Psychologists tell us that animals, including humans, have a basic need for freedom. As a result, humans have a long history of trying to live the way they want. We want to define our own boundaries. We resist boundaries placed on us by others. This is why it's natural for us not to obey imposed rules. We dislike them so much that we flirt with how far we can push them. For example, we drive our cars faster than the speed limit. Some grow or use marijuana. Many don't report all of their income to the IRS. It's basic for us to rebel against rules placed on us by others. It is also basic to push those rules as far as we can.

When young, I resisted and pushed the rules set by my parents. If they told me to be in by 10:00 p.m., I'd make it home around 10:10. If the rule said clean-up my dishes following dinner, consciously or subconsciously, I would forget more than remember. My rebellion forced my parents to constantly remind me to do chores. It's the same with my children. It was probably true with you and with your children too. We resist rules placed on us by others even if we agree with the need for them. Odd creatures, aren't we?

Why are we talking about rules in a book on total quality? Is this like the kitchen sink? While we're talking, why not throw that in too? No, not at all. If there is one single issue that separates management from labor, safety from manufacturing, traditional approaches from those of total quality, worker participation from isolation, it is the existence and application of safety rules in the workplace. If we try to fix our program or organization using total quality and do not evaluate this historical segregator, we'll miss the boat. Whatever concepts or tools we use from total quality, it comes down to how it affects the worker, doesn't it? So, let's talk about safety rules.

One of the greatest frustrations in any work environment, from both a safety and line management perspective, is the failure of employees to follow safety and health rules. "Why aren't you wearing your safety glasses? You know we require them." "Operating a vehicle requires that you use the safety belt! Why aren't you using it?" More than any other area, safety rules are the most commonly forgotten. Why is this? Aside from it being natural to buck rules, work rules have always been considered "their rules" and not "my rules." Thus, ownership in them by management and workers is limited. Hence, it's natural to resist, forget, and push.

ENFORCEMENT AND THE "VIOLATION"

Not conforming to safety rules is known as a violation. A violation is countered with enforcement. Within the traditional organization that is serious about compliance with safety rules, there is a recognizable enforcement approach. Considering safety rules as non-negotiable parts of their job, employees meet a "my way or the highway" unbending management stance. Deviations are not tolerated. Ideally, discipline is swift and firm. Management feels that it is better to err than to come across as soft or unsure when enforcing a safety rule. Management sees safety rules as "black and white." They are to be obeyed, not questioned. Unsafe actions are merely deviations from accepted safety rules and receive the classical violation reaction, enforcement.

How do workers react to this traditional approach to enforcing safety rules? In the end, workers most often acquiesce; they need the job and the rules come with the job. Polarization between management and labor occurs. This polarization is an issue of power—management with great power and labor with little. Thus, the rules become "Safety's rules" or "management's rules," which in turn result in poor worker attitudes and lack of ownership or buy-in by workers. They also force arbitration of disputes and enforcement actions, an inevitable "us" versus "them" stance and create a very adversarial organizational culture.

Of course, there are notably different ways of looking at safety rules and enforcement within application of different management theories. For example, under Theory Y management, the "huggers and kissers," safety rules become so soft that they might be best termed, "Negotiable Safety Wishes." Japanese management, however, uses management direction with team consensus for interpreting safety rules. Safety rules become equal with operating procedures—all must have compliance to reduce wastes, including injuries. The Japanese have another interesting concept about employment that speaks of this cultural safety rule philosophy. They believe that neither you nor the employee really know what's going on until the employee has worked for seven years. Only then can they be totally indoctrinated into the organization's culture.

WHAT IS THE PURPOSE OF RULES?

In our discussion of safety rules, we must ask some basic questions. What is the purpose of safety rules? The answer is the foundation of the problems concerning safety rules. Too often we only focus on the "result" and not on the "intent" of safety rules. For example, one confused result of safety rules is the attempt to gain compliance, both in a regulatory sense and in a worker behavioral sense. The word compliance becomes synonymous with the word rule. This is often associated with autocratic or Theory X management organizations. On the other hand, in Theory Y organizations, the confused result is softer—to keep workers safe. Unfortunately, this result is neither definable nor quantifiable. No wonder this approach doesn't work. One of the most commonly confused results is to avoid the costs of noncompliance. It's a back door approach to compliance. Costs of noncompliance usually include penalties, notoriety, direct accident costs, and indirect costs. We confuse that result of having safety rules with the original intent. This causes us to set rules in concrete and not question them.

Why have safety rules in the first place? Here we look beyond the result and begin to focus on intent. Safety rules are used for three reasons. First, rules are there to change behavior. The thought process is that if you require something to be done a specific way enough times, it becomes engraved as part of natural behavior. It's like the body developing a reflex arc. The second reason is to keep the brain in gear, focused on the job. It's like motor vehicle laws. If you know that the speed limit is 55 mph and signs constantly remind you of that rule, you become more conscious of your speed and watch your speedometer more. The third and main reason for rules is to produce a consistent product or result. Rules produce consistency. Most of the time, rules do just that. You can see this on our highways, standing in a ticket or hot dog line, student dress at school, and in our workplaces. Rules produce consistency.

Is that the purpose of safety and health rules? To develop industrial robots? By implementing rules, make everyone perform exactly the same? Makes you stop and think. Let's ask some harder questions. If this is true, how can we ever expect our employees to develop new and better ways of doing things? How can we get compliance and creativity at the same time? How can we ever get better at providing a safe and healthful workplace?

Obviously, we need to challenge the way we think about rules. It's another paradigm issue. Let's begin with a simple question. Are safety rules dynamic or static? OSHA and traditional management concepts would have us believe that they are static, but are they? What are the paradigms that guide this way of thinking about rules? Have you ever heard, "If it ain't broken, don't fix it?" How about, "Tried and true." Have you heard that one? "It's always been that way," is one of my favorites. These common sayings state our beliefs and lie at the foundation of our paradigms about rules. Think about it. In your family, do you operate on the same or near the same rules that you did with your parents? These paradigms not only affect our way of thinking at work, they affect every aspect of our lives.

THE "TRUTHS" ABOUT RULES

As a first challenge to the way we think about rules, let's introduce some truths about rules. First, rules reflect the workplace where they are used. Therefore, they are linked to changes in the workplace. Just as the workplace is dynamic and changes, so must our rules. Rules are linked to changes in production, processes, manpower levels, and work area or environment. As changes occur, so must the rules or their application. For example, if we eliminate overhead work, hard hats may no longer be needed. If we enclose a noisy process and reduce noise level, personal hearing protection may no longer be necessary. If we replace a toxic chemical with a much less toxic one, respiratory protection may no longer be necessary. Rules are linked to changes in the workplace.

Rules also must be linked to changes in the workforce. For example, if the workforce begins to include those that understand little or no English, we must change the way we word and convey rules. If the mix of workers changes so that shorter workers enter the work force, we may need to modify the rules. For example, women may be more susceptible to particular hazards or toxins than men. Therefore, changes in the workforce may cause changes in safety and health rules.

Too often safety rules remain static in a dynamic workplace. They don't change with the new work environment. To illustrate this point, let me tell you two stories. A husband was watching his wife prepare a roast beef one night. As she got ready to put the roast into the cooking pan, she took out a knife and cut off about one half inch from both ends of the roast. She then proceeded to place the bulk of the roast into the pan and place it in the oven. "Why did you do that?" he asked. "Do what?" she asked. "Why did you cut off a little from each end and throw the meat away?" he continued. "I don't know. That's the way my mother taught me to prepare a roast. It's the way I've always done it." she answered. But his question caused her to question the practice also. Later that week, the wife asked her mother, "Why do we cut off the ends of a roast beef before we cook it?" "Gee, now that you ask, I don't know. It's the way my mother taught me to prepare a roast," her mother answered. "Next time I talk to her, I'll ask." And, the next time she talked to her mother, she did just that. "Why do we cut the ends of a roast beef off before we cook it?" "I don't know why you do it," her mother began. "I always had to because my cooking pan was too small for the whole roast to fit into."

I got involved in a labor-management situation at its emotional peak. Both sides were horn-locked and immovable. Emotions were high. Neither side was backing down. What was the issue? A worker had been disciplined for not wearing his hard hat that, of course, was a violation of the safety rules. Labor's position was that it was a stupid rule. Management saw it as a violation only, the validity of the rule wasn't important. Being an outsider, I wasn't a part of either side's issue. I came to fix it before it got any uglier. The safety rule clearly stated that employees were to wear hard hats in the area. There was more at stake here than just the violation, however. Management credibility was much more at risk. Looking at the situation, I couldn't escape one question, "Why were hard hats required in this area?" Surveying the area revealed no overhead hazards. Talking to old timers in the shop area made the picture clearer. Years earlier they machined heavy parts in these shop areas. More than 20 years earlier, the machine shop had been converted from a heavy product machining area. With the area conversion, the jib-mounted hoists that had necessitated the hard hat requirement were removed. The workplace changed, the rule didn't—dynamic workplace, static rules. The discipline issue could have been avoided if the rules had been dynamic too and if someone had recognized the need to change them.

DERIVING AND CHANGING RULES

Let's shift gears. How are rules derived and how should they be derived? This is the second challenge to the way we think about rules. Traditionally, safety rules have one of three origins. They are developed and imposed by management, developed by "someone else" (which is lost to history or by "teflon people"), or they just appeared at some point in the past. The message is that there is no control or say so concerning safety rules. This makes the process itself bigger than life. It's like, in some countries, why do they drive on the right side of the road and others on the left? Is it biblical, a law of nature, or just as we tell our kids, "because"?

Paradigm Shift #1: In an ideal world, how should we derive rules? Imagine that we are all magically transported to a perfect workplace. In that workplace, rules are derived and evaluated by those who are affected by them. There must, of course, be "reference points." These include standards or guidelines such as OSHA, NFPA, ANSI, etc. They also include customer specifications and process guidelines. There is also room for historical information that we learn "the hard way." This include both process and chemical hazard information.

In this perfect workplace, who derives the safety rules? Those who are affected by them. Those fall into three categories: management, workers, and support personnel. Management evaluates concerns that includes providing the necessary protection, and also standardization, manageability, and enforcement. Worker concerns include having the knowledge and understanding of the hazards and precautions, having reasonable and doable expectations placed on them, and ownership and consensus in the decisions. There should also be input and consideration by support personnel including engineering, safety, environmental, quality, customer services, and human resources.

Paradigm Shift #2: When should we evaluate or change safety rules? In our perfect workplace, rules are *not* set in cement or chiseled in stone. What then is used to trigger evaluation or change? Let's answer this question with another. How often do changes occur in the workplace? Change is the trigger. Evaluating the rules is part of the management of change.

Why is it important to include everyone in the discussion about safety rules? As Coonradt puts it, with any directed activity we do, we need two critical pieces of information. They answer the questions: What and Why? We are programmed with the need for these two pieces of information before we can act. What's important about these two terms is this. People, including workers, can tell how important they are by the way we talk to them. The inescapable correlation is that we tell important people why. We tell unimportant people only what. For example, your boss always needs to know why before he can support an action or something else. In fact, we insist on telling him or her why. It portrays how smart we are and how thorough we've been. On the other hand, we seldom tell unimportant people why. We only tell them what to do. "Clean up your room." "Go to bed." "Don't wipe your nose on your sleeve." People can tell how important they are by the way we talk to them.

How does this apply to involvement in the safety and health rule process? Think about it. Do we insist upon telling workers why we need rules and include them in the process. Or, do we simply tell them what we require? Understanding why and involvement in the rule process builds "buy in." It no longer is "your rule." It becomes "our rule." A big difference.

The evaluation and development processes for safety rules provide excellent opportunities for control programs. Examples of control programs you can use include flow sheets, check sheets, and step-by-step decision guides. Ideally, these apply to this evaluation and development process. What does using control programs in this way provide us? First, it assures the use of our "reference points." Second, it forces standardization in the evaluation process. Third, it allows "nonexpert" evaluation. So, the process can be done with 100 percent confidence at the work level, not in the safety department or in the manager's office. And fourth, it entrenches the safety responsibilities in the line structure where they belong. Control programs are ideal for this use.

ENFORCEMENT OF SAFETY AND HEALTH RULES

Let's take a new look at enforcement of safety and health rules. It's strange, but something different starts to happen when workers begin to have ownership in safety and health rules. They get a lot more serious about them. Because of that, less management correction of "violations" is required. Still, even in the best organizations, some will "buck the system" just because there are rules. Also we, being human, forget at times. These still require corrective actions.

Something else happens in this empowered workplace. Given a system of peer observation and correction through anonymous observation, workers take care of their own. A mechanism such as the DuPont STOP™ program works ideally in this workplace. With any observation and correction program, however, a paper trail is still necessary. This is especially true in a workplace where everyone is responsible for identifying and correcting hazards and behavior. As workers identify hazards, the paper trail leads to management for evaluation and, if necessary, correction or allocation of resources. These management actions may include increasing or providing specialized training, raising the awareness level of the entire team to specific common issues and hazards, increasing management observation of hazards or behavior, or increasing enforcement.

ENFORCEMENT OF SPECIAL CASES

Two issues that are separate from routine observation and correction of hazards and behavior. These would include repeated violations and cardinal rules. Cardinal rules are those that, if violated, may result in fatalities. These include certain hazards like confined spaces, lockout, and live electrical work, to name a few of the more common ones. We can't handle repeated violations and those involving cardinal rules in the same manner as lesser behavioral problems. The risks aren't the same. Management must deal with these specifically, quickly, and firmly in a standardized manner.

Safety and health rules. Very often they are the glue that keeps the safety and health program together and workers safe and healthy. In a total quality program, one cannot discount the importance of evaluating how we derive, implement, and enforce safety rules. There must be new thinking in this area if it is to dovetail into total quality. Rules must be dynamic, just like the workplaces, workforces, and processes. Everyone affected by the rules must have input into the evaluation and definition process. Enforcement must also be re-thought in an empowered workforce. In a total quality workplace, safety and health rules aren't factors that segregate safety from management or management from workers. They are tools that build participation and a safer and healthier work environment.

15

BENCHMARKING AND
PROGRAM EVALUATIONS

There are many complaints about benchmarking. "It just doesn't work." "It's too confusing." "There isn't enough common ground to work." "I think benchmarking is overrated. It's just another buzzword going around today." It's true, benchmarking is a common buzzword that organizations and managers are using today. It comes from the "everyone's-using-it" part of our vocabulary. You know, trendy stuff. But, benchmarking really isn't trendy, and it isn't complex. It's confusing because of the multiple ways that it is used by those who either aren't benchmarking or are using the term incorrectly. Is it overrated? No, it is simply misunderstood and misapplied.

When you talk about program evaluations, one word always comes to mind—auditing. Auditing brings a negative picture to our mind. Auditing is that detailed, point-by-point, word-by-word, step-by-step process that looks for picky little areas that really don't matter at all. Auditing is that adversarial, confrontational process like when the IRS audits your tax return and records. Are program evaluations really just another way of saying auditing? No, but it's no wonder that we get defensive when someone suggests that we evaluate our program. "Trying to find out what I'm doing wrong, huh?"

These two terms, benchmarking and program evaluation, conjure up confusion, misconceptions, negative thoughts, and fearful emotions. In reality, however, they are very positive concepts that go hand-in-hand. Look at them as being different in only one aspect. One looks outward, the other inward. They are both total quality processes that aim toward program improvement.

BENCHMARKING

Let's take a closer look at benchmarking first. It might be easier (or less confrontational) to look at an outwardly focused concept before we look inward. What is benchmarking? Benchmarking is simply learning what other programs and organizations are doing to be successful and using some of those successes in our program or organization. Benchmarking is an effort to prevent you from having to "reinvent the wheel."

Many think benchmarking is a new concept. Not so. It's as old as competition. In the Middle Ages, when one craftsman began blowing glassware with a handle, all the rest did too. As one blacksmith began to forge his heated metal to make a harder plow head, the other smiths followed. When they introduced the first diet soda, the other brands quickly followed. A runner uses a very successful training method and other runners copy it. When a new leaner-meat hamburger or a light beer goes on the market, the other competitors are right behind with their own products. In business, we call benchmarking "competition-induced product development." In safety and health, however, we're really not in competition with other safety and health programs are we? So, the reason for adopting successes changes when there's no competition. We don't benchmark to keep up with the competition. We benchmark to elevate the safety and health mission, to get better at providing a safe and healthful workplace. From a less altruistic vantage, benchmarking can make our jobs easier and more successful.

Before we discuss the specifics, however, it's a good time to point out the number one reason benchmarking goes astray and negative feelings develop. It is very common to get caught up in the concept and rush into benchmarking before having your program's "ducks in a row." Many begin focusing on outside organizations or programs before solving most of their own problems. This causes immediate frustration and a lot of wasted effort. In common cases like this, conflict occurs on two fronts: 1) what is already deficient or weak in their program and 2) what can they import. It just doesn't work. You can't "serve two masters." There is a simple, before you start, rule. It says, you first need to slay the dragons that are bringing your program down before you look outside for additional improvement ideas.

As with any concept, it's the process that's important to us. The benchmarking process usually contains six parts.

Parts of Benchmarking

1. Survey
2. Identify
3. Prioritize
4. Develop
5. Implement
6. Follow-up

Benchmarking Part 1: The first part in benchmarking is *surveying* "front running" programs or organizations. This is where a lot of benchmarking efforts go astray. Why? First, there is a critical need to identify the programs or organizations that are truly front-running. Too many times due to expediency, ignorance or laziness, little time or effort is spent studying programs or organizations to determine which ones are at the front of the pack. Poor choice of the benchmarking targets will poison the process before it ever gets off the ground. The second reason why benchmarking efforts go astray is that good programs or organizations are chosen but significantly different ones aren't excluded. For example, a human resources benchmarking effort might study a unionized plant's program when they wish to apply it to a nonunionized operation. A maintenance program may benchmark in a very different industry sector. Surveying errors and benchmarking too soon account for about 90 percent of the bad benchmarking experiences and poor opinions about benchmarking.

Once front-running organizations or programs are selected, the sites are visited or telephoned to see how they operate. A survey is used to gather information and to answer these questions, "What makes this organization or program one of the best?" "What are they doing differently?" And, "How do they think differently?"

Benchmarking Part 2: The second part of benchmarking is *identifying* the complementary solutions used by the target organization or program. Complementary because the purpose of benchmarking is not to build a replica of the target. In fact, every organization works differently because of the organization's culture, management, market, and other internal and external factors. Thus, some of the aspects or solutions used by the target won't dovetail with your program.

Benchmarking Part 3: Part three of benchmarking is *prioritizing* the "growth opportunities" from the list of complementary solutions. It's a simple matching and weighing process. Where are your programs weak in comparison to theirs and what one(s) will give you the biggest return if used? Trying to bite off too much at once can also defeat the whole process. An ambitious benchmarking effort will only look at one or two growth opportunities at a time.

Benchmarking Part 4: *Develop* a plan to achieve growth. Which growth opportunities will you pursue? In what order will you pursue them? What time frame will you use? Who will be responsible for each item? The chapter on planning should be helpful here. Realize, of course, that having identified growth opportunities is not enough. Without a specific and detailed plan for implementing those opportunities, the efforts will be inefficient and, for the most part, wasted.

Benchmarking Part 5: *Implement* the plan. Put the power to the paper. Don't make a plan and then put it away. Apply the resources to what you have planned, track the plan regularly, and bring it to completion.

Benchmarking Part 6: Benchmarking is a dynamic process. Once you've implemented your plan, you're not done. This sixth step is the on-going aspect of the process, the *follow-up*. At regular, scheduled intervals, resurvey, reprioritize, refine the plan, and redirect the implementation. This portion fits nicely within an annual strategic planning process.

What value is benchmarking? The most obvious is that benchmarking provides a comparison of your program to "front runners." It's a measuring stick of how good a program is. Second, benchmarking allows learning from successes. It's like opening the window to freshen the air. Too often we get myopic. It is common to become drunk on our own existence, thinking that we've made it into that upper circle. Actually, we lose touch with what is really going on. We become the reference point, not the true front runners. Third, benchmarking helps prioritize improvements. This not only helps focus your improvement program, but it avoids the confusion and frustration that goes with trying to improve everything at once. And fourth, benchmarking provides a mechanism for continual growth. This is a very valuable aspect of benchmarking. It's a continuum toward achieving and maintaining excellence in your program or organization.

BENCHMARKING STEPS

How do you benchmark? Generally speaking, it's a step-by-step process that covers all the benchmarking parts. Here is a step-by-step approach to benchmarking. A word to the wise—if you are beginning a benchmarking effort, don't skip the first step.

Benchmarking Steps

1. Sell the program and get permission/resources from upper management
2. Identify the front running programs/organizations
3. Arrange for the visit(s) or call(s), visit or call the facility or facilities, and evaluate their successes
4. Pool all the information
5. Identify program/organizational areas for improvement by comparing yours to theirs
6. Prioritize the improvement areas
7. Develop an improvement plan
8. Implement the plan
9. At regular intervals, evaluate the progress/plan

PROGRAM EVALUATIONS

We shift gears here. Benchmarking is looking outward for ways to improve. Program evaluations look inward. They must be a precursor to the benchmarking process. Without knowing where you can improve, it is nearly impossible to prioritize improvements. Also, evaluating your program before you visit a target program or organization helps focus your benchmarking efforts and identify what information you need. Program evaluations are not negative processes although they are very often thought to be. Program evaluations, properly used, are extremely valuable *and* positive.

For simplicity sake, our discussion on program evaluations is divided into three areas: customer evaluations, self-evaluations, and trend analysis.

CUSTOMER EVALUATIONS

Customer evaluations are commonly known as quality surveys. They determine how you are meeting your customers' needs and expectations, and what your customers' perception is of the quality and value of your services or products. Conduct a reality-check before doing *any* customer evaluation: "If you don't want to know, don't ask." This thought goes deeper, though. "If you will discount what your customers say or won't be willing to do anything about it, why ask in the first place?" This is one of the major problem areas. It's like having your spouse or boss tell you what you could be doing better. It's natural to resist and become defensive. "What do they know?" Evaluations considered in a positive light—seeing what opportunities are there to strengthen a program rather than what is being done wrong—meet with much less defense and emotion. If you feel emotional or defensive about asking customers what they think, you are in good company. The real devil, however, is in the perception of the evaluation, *not* in the evaluation process itself.

Customer evaluations can have two different focuses: external customers and internal customers. Generally, external customers are those who are not onsite, while internal ones are. To a safety and health program, external customers might include clients (consulting business), corporate departments and services, regulatory agencies, etc. Internal customers might include human resources, supervisors and managers, workers, trainers or training, engineering, communications, etc. A lot of confusion and emotion surround whether people or groups are customers or stakeholders. It's a terminology game. Don't get caught up in this trivial point. Evaluations seek perceptions in the first place, not facts. This being the case, what difference could it possibly make if they are stakeholders or customers?

With both external and internal customers, two kinds of surveys, soft surveys and hard surveys, can be used. Soft surveys are, at best, semi-quantitative in nature. They ask customers to "rate" performance or needs on some type of scale. It is only semi-quantitative because it is totally subjective in measurement, and therefore, based on soft data. It's called soft data because it is a measure of perceptions. Hard surveys, however, ask specific questions and seek detailed information such as, "What could be added to improve the service we provide?" Hard surveys are more fill-in-the-blank and essay-type than scaleable. Both kinds of surveys, soft and hard, provide specific advantages. Soft surveys give good trend data to single-point data. Hard surveys are very specific and provide "hard" information.

SELF-EVALUATIONS

Self-evaluations usually follow a specific format. Annually, many use program-specific self-evaluations. For example, in a safety and health program self-evaluation, program specifics might include industrial hygiene, safety administration, motivation, communicating and reporting, hazard control, etc. As an example, Westinghouse Electric Corporation uses a corporate-wide self-evaluation format. Because it's a good example, it's included in Figure 5 at the end of this chapter. Self-evaluations use a semi-quantitative rating system that is based on performance. They are best done by team. This levels the playing field by reducing the possibility of biases or knowledge levels affecting the ratings. Like the soft customer surveys, self-evaluations are best used for trend analysis and to determine what areas have slipped or gained from one evaluation period to the next.

Customer evaluations and self-evaluations are extremely valuable to the continuous improvement efforts of a total quality program. They provide information from outside the function or group and outside the myopia of the function. As many in public relations have said for years, "Perception *is* reality." Customer evaluations provide this sometimes painful look at that reality. Too often, we get into the "can't see the forest for the trees" situation and miss what we are doing well or what we can improve. Self-evaluations provide this information very well.

Customer evaluations and self-evaluations are dynamic processes. Left alone as a static art, they can be frustrating, demoralizing, and provide little true information. In other words, depending on what you ask, how you ask it, who you ask, when you ask it, and the mood of the target group when you ask, you can add so much bias to the evaluation that it can tell you the exact opposite of reality. That's why evaluations *must* be dynamic processes. The greatest value is to document the process of continuous improvement. Trend/progress analysis provides this information.

TREND ANALYSIS

In trend analysis, the evaluation formats must have a quantitative scale. Many different scales are available. The biggest complication is trying to use too many points or too detailed a scale for evaluation. The best and most repeatable are those who use scales with around five levels such as 1 to 5 with one being poor, two being fair, three being OK or average, four being good, and five being excellent. These five levels are easily understood by most of us and present minimal confusion. Scales that are too large or not well defined such as 0 to 10 with zero being low and 10 being high, leave too much to interpretation and are too confusing. For example, on such a 0 to 10 scale, I might think that a rating of 5 is OK or average, while you may think that 7 is minimally acceptable because it more closely relates to 70 percent or a C grade. Thinking in this manner, you would view my 5 ranking as a flunking grade when I meant it to reflect an OK performance.

Choosing the proper scale and definition of the points are important parts of making trend analysis meaningful. After all, trend analysis that is usually an annual effort, grows stronger (more powerful) the more data points or years that are available as input. One data point tells you little; at best it is a reference point. Two data points are interesting but also may tell you nothing. Two points give you little more than a line when connected. Three data points begin to say something about whether the performance is static, going down, or improving. With three, you begin to get a trend. When you get to four data points, or four years of evaluation information, you start to get valuable (powerful) data from your trend analysis. And, each year following, the trend and data will be more and more valuable and dependable.

This is a major reason for developing your evaluations carefully. They must stand the test of time. If after one or two years you decide that the evaluation doesn't ask what you want, needs to ask more questions, or the scale isn't useful, you *start back at zero* with your trend/progress analysis!

Benchmarking and program evaluations are fine, but you must understand that they are tools for continuous improvement. Therefore, whenever you use benchmark or program evaluations, it can't just go into the file for next year's efforts. It is critical to develop action plans from the results. What program elements need improvement? What ideas are there to improve your program? Develop an action plan and track the plan throughout the year to completion. Then, in the next benchmarking or evaluation effort, you can see if the results are a function of your action plan. Did you improve because of what *you* did or what *they* did?

PARTNERSHIPPING

Complementary to this continuous improvement process is the use of partnershipping opportunities. Partnershipping is a form of bridge building, tapping team synergy for improvement. A significant and helpful one is the OSHA Voluntary Protection Program (VPP). This program provides a valuable self-evaluation process and an external evaluation of your safety and health program. It does, however, notably challenge some of industry's most strongly held paradigms—those paradigms concerning OSHA being an adversary and not a partner in worker safety and health. It also challenges paradigms held by organized labor. Primarily those about changing from a separated relationship with OSHA to a partnership. This paradigm fears that partnershipping reduces the effectiveness of OSHA as an outside, open-minded, advocate for worker safety and health. These paradigms from the past just aren't true. In fact, through this partnershipping, you can develop better safety and health programs and worker safety and health can be improved.

Another helpful partnershipping opportunity is industry and business groups such as the Chemical Manufacturers Association, area business organizations, and manufacturers organizations. These opportunities provide valuable opportunities for exchanging ideas and sharing successes. Too often, however, these organizations are only viewed as important to the production or engineering folks. Safety and health stand on the outside and don't receive the advantages that involvement in these organizations and associations can provide to benchmarking and program evaluation efforts.

The last area of partnershipping opportunities is one that safety and health professionals are well aware of, but seldom use effectively. That area is involvement in safety and health professional associations and societies. Too often professional societies are only looked at as certification maintenance or social opportunities. The networking opportunities that they provide can give significant advantages to safety and health program benchmarking and evaluation efforts.

TOOLS TO IMPROVE YOUR PROGRAM

This chapter has looked at improvement, specifically total quality tools that can help you improve your safety and health program. Let's make one more important point. Benchmarking, program evaluations, trend analysis, and partnershipping are valuable tools that can be used to improve a program. They are parts of an effective total quality program. Most programs already have a minor form of this called the annual performance evaluation or review that is a part of the salary review process. The major problems, of course, are that these are subjective and are done by someone who usually doesn't understand the entire picture. Performance is also most often measured using only observation or judgment. What the boss or evaluator sees or perceives is how the ratings are determined. Benchmarking and self-evaluations provide mechanisms that document the quality and progress of any program. Used in the annual performance management system, they can provide an evaluation of performance by measurement, *not* solely by judgment or observation. This benefit alone should convince all safety and health professionals to get involved in these improvement opportunities.

11.3 Rating Sheets

S&H PROGRAM EVALUATION

The numerical values to be inserted below are the weighted ratings calculated from the attached evaluation sheets. The total of these ratings becomes the overall rating for the facility.

			Previous Yr.	Current Yr.
I.	Management Commitment	=	_____	_____
II.	Employee Involvement	=	_____	_____
III.	Workplace Analysis	=	_____	_____
IV.	Hazard Prevention & Control	=	_____	_____
V.	Training	=	_____	_____
	Overall Facility Rating Total	=	_____	_____

Facility Experience

	Previous Years		Current Yr.
	19_	19_	19_
* Total Recordable Case Rate			
**Lost/Restricted Workday Case Rate			
***Lost Workday Rate			

(*Lost workday cases + non-lost workday cases) x 200,000 = Total Recordable
    Total Hours Worked   Case Rate

**(Lost/restricted workday cases) x 200,000 = Lost/Restricted Workday
    Total Hours Worked   Case Rate

***Lost workdays x 200,000   = Lost Workdays
    Total Hours Worked   Rate

Facility_____ Date Appraisal Made_____

Evaluation Participants_____
              _____
              _____
              _____
              _____
              Submitted by_____

Note: Score each element by circling number located under appropriate rating heading.

I. MANAGEMENT COMMITMENT

	None	Fair	Good	Excellent	Comments on "None" and Fair Items
1. Statement of a site policy on the S&H and healthful working conditions.	0	5	10	12	
2. Established, clear goal(s) for the S&H program and objectives for meeting the goal(s)	0	5	10	12	
3. Visible top management involvement	0	5	10	12	
4. Site S&H committee(s), meeting preparation and follow-up	0	3	6	8	
5. Special S&H task force committees	0	2	5	6	
6. S&H staff	0	2	5	6	
7. Site S&H manual	0	2	5	6	
8. OSHA regulations & standards, reference material	0	2	4	5	
9. Responsibility for all aspects of the S&H program are assigned and communicated	0	2	5	6	
10. Adequate authority and resources provided to responsible parties	0	3	6	7	

Note: Score each element by circling number located under appropriate rating heading.

I. MANAGEMENT COMMITMENT (Cont'd.)

	None	Fair	Good	Excellent	Comments on "None" and "Fair" Items
11. Managers, supervisors, and employees held accountable for meeting their responsibilities	0	4	8	10	
12. Annual review of the S&H program	0	4	8	10	

Total Value of Circled_____ _____ _____ _____ X .25 Rating
Numbers

Note: Score each element by circling number located under appropriate rating heading.

II. EMPLOYEE INVOLVEMENT

	None	Fair	Good	Excellent	Comments on "None" and "Fair" Items
13. Provisions and encouragement for employee involvement	0	6	12	15	
14. Safety Observer or equivalent program involving employees	0	6	12	15	
15. Site S&H committee	0	6	12	15	
16. S&H task teams	0	6	12	15	
17. S&H inspections	0	4	8	10	
18. S&H injury and illness investigations	0	4	8	10	
19. S&H meetings	0	3	6	8	
20. S&H promotion, communication, information	0	3	5	7	
21. Off-the-job S&H program	0	2	4	5	

Total Value of Circled_____   _____   _____   _____ X .20 Rating
Numbers

Note: Score each element by circling number located under appropriate rating heading.

III. Worksite Analysis

	None	Fair	Good	Excellent	Comments on "None" and "Fair" Items
22. Baseline Safety and Hazard Worksite Assessments	0	6	11	14	
23. Periodic Safety and Hazard Worksite Assessments	0	6	11	14	
24. Analyze New or Altered Processes and Materials	0	5	10	12	
25. Analyze New, Altered or Relocated Equipment or Facilities	0	5	10	12	
26. Routine Job Hazard Analysis	0	6	12	14	
27. An Employee Health and S&H Concerns Program	0	4	8	10	
28. Initial Responsibility for Accident and Incident Investigations by Line Supervision and Follow-Up by S&H Staff	0	4	8	10	
29. Analysis and Reports of Injury and Illness Trends	0	3	6	8	
30. Periodic and Special Studies	0	1	2	3	
31. Off-the-Job Injury Reporting	0	1	2	3	

Total Value of Circled _____ _____ _____ _____ X .17 Rating

Note: Score each element by circling number located under appropriate rating heading.

IV. Hazard Prevention and Control

	None	Fair	Good	Excellent	Comments on "None" and "Fair" Items
32. Physical Hazards Control - Machine Guarding, Radiant Energy, Heat, Noise, Lasers, Illumination, Etc.	0	4	8	10	
33. General Engineering Controls	0	4	8	10	
34. Hazardous Material including Flammable and Explosive Control, Storage and Labeling	0	4	8	10	
35. Ventilation	0	4	8	10	
36. Inventory of Hazardous Materials and Utilization of SPDSs and/or MSDSs and Process and Finish Specifications	0	4	8	10	
37. Material Handling- Manual and Mechanical	0	4	8	10	
38. Housekeeping - Storage of Materials, Waste Control, Etc.	0	2	4	5	
39. Personal Protective Equipment and Skin Contamination Control	0	4	8	10	
40. Medical, First-Aid and Emergency Response Program	0	4	8	10	

IV. Hazard Prevention and Control (cont'd.)

	None	Fair	Good	Excellent	Comments on "None" and "Fair" Items
41. Maintenance of Physical Equipment - Cranes, Fork Trucks, Guards, Hand Tools, Etc.	0	2	4	5	
42. S&H for Contractors	0	2	4	5	
43. Corrective Action Tracking System	0	2	4	5	

Total Value of Circled Numbers _____ _____ _____ _____ X .18 Rating

Note: Score each element by circling number located under appropriate rating heading.

V. Safety and Health Training

	None	Fair	Good	Excellent	Comments on "None" and "Fair" Items
44. Manager's S&H Training	0	4	8	10	
45. Line Supervision Basic Course and Follow-Up	0	6	12	15	
46. Indoctrination of New or Transferred Employees (Including Hazard Communication)	0	6	12	15	
47. Training for Specialized Operations - Fork Truck, Crane, Paint Spraying, Welding, Lockout/Tagout, Electrical Safety-Related Work Practice, Confined Space Entry, Respirators, First-Aid, CPR and Bloodborne Pathogens, Etc.	0	6	12	15	
48. Workplace Meetings or Tool Box Session (Employee Involvement)	0	4	8	10	
49. Vehicle Safety	0	3	6	7	
50. Safety Observers Formal Training and Follow-Up	0	4	8	10	

V. Safety and Health Training (Cont'd.)

	None	Fair	Good	Excellent	Comments on "None" and "Fair" Items
51. S&H Train-the-Trainer Program	0	3	6	8	
52. System to Document Training and Auditable Follow-Up	0	4	8	10	

Total Value of Circled Numbers _____ _____ _____ _____ X .20 Rating

16

CONTINUOUS IMPROVEMENT

A warning is in order before starting this chapter. The concepts presented in this chapter are counter to the way Americans think. Whether Americans talk about the economy, the way they look at industrial improvement, profits, children or spouses, behavior in general, or work, when things are bad, Americans always refer to "turning things around." The unsaid qualifier is that it is expected that this turnaround will happen immediately. Whether it refers to changing the performance of a subordinate, improving the nation's economy, changing the nation's health care system, fixing your car, having your child stop an annoying habit, or any other changes in our lives, Americans expect them to happen *now*! If change is necessary, "let's get on with it." Historically, Americans have *never* been known for their patience. This chapter is really about patience, patience in the process of change.

This traditional American expectation is not just an "Old World" way of thinking. Even today, business books and articles in America still teach this concept of sudden, drastic change. We have a word for it. We proudly call it innovation. It is change that is so new, so cutting-edge, and so radical that it turns processes and organizations around immediately.

KAIZEN

Masaaki Imai talks about this American way of thinking in his book *Kaizen, The Key to Japan's Competitive Success*. He describes the environment where innovation is most successful. That environment would include rapidly expanding markets, consumers oriented more toward quantity rather than quality, abundant and low-cost resources, a belief that success with innovative products could offset sluggish performance in traditional operations, and management more concerned with increasing sales than with reducing costs. That may have been the way it used to be in America, but is this the world in which we live today? Does this describe the drivers in the world economy now? No, everyone knows that it is very different. Many of our manufacturing and service providers, however, *still* operate as if they believe this environment for innovation still exists. Any wonder why America's industrial strength is suffering?

How has this environment changed? In short, it's changed 180 degrees. Today's business environment would include a sharp increase in the costs of materials, energy, and labor. It would include overcapacity of production facilities, increased competition among companies in saturated or dwindling markets, changing consumer values and more exacting quality requirements, a need to introduce products more rapidly, and a need to lower the breakeven point. This is the environment where continuous improvement flourishes and innovation falls way short.

It would be natural to think that this is a chapter that addresses business only and applies little to safety and health programs. But, draw direct correlations to inner-organizational environments that have a direct and magnified effect on your program. How have the sharp increases in the costs of materials, energy, and labor affected your safety and health program or manpower? How does operating at lower and lower production rates, under the design limits of the facility, affect your safety and health program, your manpower, or add challenges to your job? Has competition between departments or staff functions for dwindling resources increased in your organization? Is there a changing expectation in upper and middle management about the value and quality of the safety and health service? Has demand increased to implement safety and health programs and corrective strategies quickly? Is there more and more demand to control hazards with less money because of the tight budget problems within your organization or department? Think about it. We aren't addressing just changes to business when we talk about continuous improvement. We are addressing changes that are universal to both your organization *and* your function. The environment has changed. And the fact is that if you *and* your organization are to be successful, you also must change. The process of continuous improvement can provide the tool for doing so.

Let me draw from a quotation that Masaaki Imai used. He quoted an American business executive who talked about this changing environment. Picture yourself and your safety and health program in his words. The American business executive told of what his company's chairman had said to top management, "Gentlemen, our job is to manage change. If we fail, we must change management." The executive added, "We all got the message!"

FOCUSING ON RESULTS *VERSUS* THE PROCESS

As you might have guessed, there's a fundamental difference between the American concept of rapid, innovative change and the total quality, or Japanese, concept of continuous improvement. The difference has to do with the focus of the process of change itself. The American way is results-oriented. Continuous improvement is process-oriented. These are *very* different. As safety and health practitioners, we are very familiar with this American results-oriented approach. What is the prime indicator that we have used to judge our success or failure? If you're like 99.995 percent of safety and health folk, you've used the injury/illness rates and lost work day rates haven't you? These are results, not processes. More safety and health practitioners than not have become frustrated chasing the objective of turning the rates around. This is the wrong focus. It *is*, however, totally American in its thinking.

Continuous improvement, on the other hand, is process-oriented. By focusing on the process, manufacturing or service related, one can "fine tune" the process, a little at a time, and make it work better. The bottom line is that the desired results will eventually happen, not overnight, but they will happen. And when they happen as a result of a process-oriented approach, they stay there. The up and down, success and failure, good year and bad year cycles are eliminated. Why? Because, the process itself is improved and the process continues.

Too often, we focus only on the end result. We try anything that will cause a quick change. It's called "the quick fix," "the plan of the week," and many more uncomplimentary phrases. These uncomplimentary descriptions, however, speak very eloquently of the success rates, don't they? Statistics bear out that in companies, organizations, and functions including safety and health, quick fix strategies have a very low probability of success. Look at the Japanese. The continuous improvement, process-oriented strategy obviously does work. The important message is that it doesn't just work in the manufacturing sector—it works in any service or staff function including safety and health programs.

SMALL AND CHEAP

Continuous improvement comes from the concept of *kaizen*, the Japanese word for continuous improvement. It focuses on small improvements, not the American way of making massive and innovative improvements. As an example, for a long time in America, we have equated improvement with the word automation. Automation is that re-engineering process where entire production facilities change to expensive, computer-controlled robotic processes where the need for workers is much lower. This isn't the focus of kaizen. Japanese use the word *jidohka,* or autonomation, to describe a different process. It's a process that works *with* people. Instead of automation, which makes more products faster, autonomation makes existing machinery or processes stop automatically whenever a defective product is made. The focus is not on making more parts or products, but rather on not making defective ones.

This doesn't involve massive change or necessarily use automation. It uses a lot of small improvements over a long time.

The beginning of any improvement has to be the point at which you recognize that there is a need to improve. Sounds pretty basic, doesn't it? It's the level of that recognition, however, that's important. Let's revisit a point made in an earlier chapter. There are four triggers for change: opportunity, recognizing a need, discomfort, and pain. The farther the trigger for change is driven back in the order, the less traumatic, more subtle, and less disruptive the change will be. This is the focus of kaizen. Too often, especially in safety and health issues, the improvement or change trigger only comes when the "wheels come off," and the injury rates go way up. This is pretty near the pain level. Kaizen moves that trigger for improvement back to the recognized need level, far before it becomes painful or uncomfortable. Only at that level can small, incremental changes be made to improve the program and make it better.

Another aspect of kaizen is that it prefers to use inexpensive or no-cost changes, rather than expensive ones. Too often we waste valuable time waiting for available funds to make large, expensive changes. And once implemented, they often don't work. We have a very poor record across industry in succeeding at implementing expensive changes. Kaizen uses inexpensive and sometimes no-cost changes that can usually be fit into an operating budget. Changes such as redirecting a product flow, moving a conveyor, changing work position, and dropping a redundant approval step in a process are good examples of this focus.

Obviously, kaizen requires a very high level of employee participation and authorship. In a total quality program, workers, not management or engineering, author most changes. This is very different from traditional American concepts about the origin of great ideas.

OPPORTUNITIES FOR KAIZEN

What tools can you use to implement kaizen, the process of continuous improvement, in your safety and health program? Let me offer seven tools or opportunities for the attentive safety and health practitioner.

First, ever hear of an *employee suggestion program*? Real new technology, isn't it? What are your first impressions of an employee suggestion program? Do you picture a suggestion box that attracts dust? Do you think of it as an employee complaint pathway? Do you equate it with a lot of crazy ideas that aren't practical? Most first impressions and, unfortunately, lasting impressions, are negative about employee suggestion programs. Why is that? It's because of the reason employee suggestion programs start and the way they are implemented. Most often, employee suggestion programs start as a tool to let employees release some steam or to give the impression that management wants the ideas, when, in fact, they don't. Traditionally, how does management view these suggestions? They see little value in them. How do employees feel about the "Box"? They aren't blind or deaf. They can "hear" the spoken and unspoken responses from management. They see the low implementation rate of their ideas. They see the suggestion box getting opened less, if at all. Most employee suggestion programs don't work well and have poor reputations because of why they start and how they are implemented.

Effective employee suggestion programs do exist. Good employee suggestion programs have some common and critical elements. First, they are dynamic. Second, management emphasizes the program in an upbeat manner. Third, the employee has a large or sole part in the ideas implementation (empowerment and authorship). Fourth, the program has high management support. Fifth, a formal recognition program is established around the program. Sixth, an award program is tied to successful ideas. And seventh, a high percentage of successful ideas are submitted. If your organization has an employee suggestion program, look at it. Does it have these elements? If it doesn't, how can you improve that program?

The second tool you can use to implement kaizen is the *line stop* power of Just-in-Time. Do you remember the purpose of line stop? It provides immediate recognition of a problem, and the interruption provides a total focus on solving it. Line stops aren't just for production or quality purposes. They must apply to safety and health purposes also. Who has the line stop powers for safety issues in your organization? Do you as the safety or health person or just upper management? The line stop powers must be driven down to *everyone* in the organization, including the newest hourly employee. This is critical if kaizen is to be effective at identifying and solving safety and health concerns.

We've already talked about the third tool—benchmarking and program evaluations in Chapter 15.

The fourth tool you can use is *process mapping*. Process mapping focuses directly at kaizen. Process mapping is a team activity where a process, including safety and health program issues, is dissected into its component steps and times. It can be, and usually is, a laborious task. When complete, however, the inefficiencies and problems are easy to identify *and* eliminate. It isn't a fast process, but it is thorough. Process mapping can be a valuable tool for implementing kaizen in any program or service.

You might not think of it, but *quality problems* can be an important tool for incorporating kaizen into a safety and health program. Production, quality, safety, costs, or morale are not isolated, stand-alone issues. When problems occur in one of these, they usually occur in all. When a quality problem happens, either a process or system has failed. What better time to review safety and health issues and implement hazard control strategies? Safety cannot operate effectively in any organization isolated from other issues; there must be active internal program partnershipping to work effectively.

The sixth tool you can use to implement kaizen is in the area of *wastes*. Taking from the Just-in-Time concepts, the goal is to identify and minimize or eliminate wastes. These are opportunities for kaizen, and safety and health must play an active role. Wastes are also ideal opportunities for merging continuous improvement in safety and health programs. The trick is the trigger. The lower the trigger toward the recognition of need level, the greater the opportunity to improve all of the processes.

The last tool is *high inventory*. Wherever we identify high inventories, there are ample opportunities to implement kaizen in your safety and health program. Whether it is raw materials, in-process inventories, or finished product inventories, operating at too high of levels is wasteful of an organization's resources *and* presents hazards to eliminate.

The trick to using many of these tools and opportunities is involving yourself in the organization's efforts and communications. Kaizen, in an organization, isn't just a manufacturing or quality concept. It's an organizational concept of which safety and health are part. Look at these tools. How many are you currently using to improve your safety and health program? How can you take advantage of more of these tools? Merely by asking these questions, you take the first steps to bringing kaizen to your program.

KAIZEN: A DIFFERENT PATH

Because success comes in much smaller advances or improvements using kaizen, an increased need for measurement arises. In an innovation strategy, success (or failure) is immediately clear. The big change happens or it doesn't. Too often it doesn't. In kaizen, change is much more subtle. We see improvements in small or fractions of percentages. This places enormous demands on measurement so you know how and if you're progressing. This is also true of kaizen as it is applied to safety and health programs. It requires that we improve how and what we measure so that we can see these small, incremental improvements *and* celebrate them.

Is it best to place kaizen into a safety and health program? Or, is it best that safety and health be part of the organization's continuous improvement program? The answer to both questions is yes. Kaizen has great advantages within a safety and health program. The organizational culture, however, must undergo change so the magnitude of change is appreciated by upper management. For the ultimate reward, however, both to the organization and to the safety and health of workers, kaizen should be inserted into the entire organization and become a part of the kaizen culture. Safety and health must be active leaders in that effort by crossing boundaries, actively communicating, and becoming deeply involved in the organization's kaizen effort. The safety and health practitioner's role in kaizen is, therefore, not only internal and program focused, it is very much team oriented.

Obviously, kaizen is not the first total quality concept or tool to pursue. It complements a full total quality process once you shift paradigms and change cultures. Kaizen alone, set adrift in a traditional organization, can do little more than frustrate those involved and fail.

The concepts of total quality contain no magic. Many companies have been led down the path only to discard total quality as all fluff and little substance because they haven't realized this simple fact. Total quality, however, is extremely powerful if used correctly. Once instilled in a program's or organization's culture, the concepts of total quality can become invaluable tools for tomorrow. Kaizen is a concept that connects the other total quality concepts and tools. It is also dependent upon them. Continuous improvement is the way to the future, but is only effective when the whole total quality animal is eaten. Taken by itself, it will only produce heartburn.

17

GOOD OR BAD, EVERY BUSINESS HAS A CULTURE

Culture, borrowing the words of Morris Massey, "is like a navel, we *all* have one!" Unlike navels, however, there are wide differences between the way cultures look. They can be open or closed. They can look toward or away from risks. They can look outward or inward. They can be dictatorial or participative. They can be positive or negative. They can be successful or not. They can be old or new. They can be cast in iron or changing. Don't shortchange the importance of having a personal culture. Like it or not, the way your culture looks at life, defines *everything* about you, your business, your organization, and your safety and health program.

FACTS ABOUT CULTURE

If we all have a culture, what are the important facts about it? What are the inescapable realities about cultures that we need to be aware of? It would be nice to know the facts about our culture, wouldn't it? There are seven facts concerning cultures. First, everyone, every group, every organization, every company, every business has a culture. Your safety and health program has a culture. Even *you* have a culture!

The second fact about cultures is that they are the "mindset" of that person, group, organization, business, etc. It's the collective "way of thinking" of that entity. A good analogy is computer programming. Computers don't work without programming and they work exactly as programmed.

Third, the culture defines everything for that individual or group. This includes important and seemingly unimportant aspects of everyday life. For our purpose, however, the more important organizational ones include how the organization communicates, how the parts work together, what values it has, how the organization views and deals with change, how secure it is, and what risks it is willing to take. All of these "definitions" for that organization directly affect total quality, its implementation, and success.

The fourth fact concerning culture is that we acquire it over a long period. Our culture is a long chain of many influences, actions (experiences), and information. Think of it this way. Development of your individual culture began as a baby and continually evolves from the experiences you have and the knowledge you gain every day. As you gain experience and knowledge, you learn what is successful or what fails. These pieces of information mold your culture. Changing your culture, therefore, takes a long time. Rapid 180 degree culture change isn't normal. Changing your personal or organization's culture takes new experiences and new knowledge that differ from past ones. That takes time, a lot of time.

Fifth, we protect our culture, and sixth, it isn't easy to change our culture. Cultures are very deeply ingrained. What we perceive as attacking our culture meets a strong defense. Attacks on a culture are threats to the heart of the personal or organizational beast. Thus, it isn't easy to change a culture. But it isn't impossible! It just isn't easy.

And seventh, in summary, any change to a culture takes a long time, requires a dedicated long-term effort, and is resisted. It matters little if the change is positive or not. This fact holds true, *always!*

IT'S YOUR CHOICE

You may choose three different possible results from reading this book. You can reject everything. You can take some or all of these total quality concepts and tools and use them in your safety and health program. Or you can embrace total quality and become a leader in the change efforts of your organization. The total quality concepts and tools in this book only have been "experiences" and "information" I have used in an effort to convert you to total quality. This book has really had one purpose from the beginning—changing your personal culture. Through the concepts and tools of total quality and the many examples and experiences we've shared, I have sought to make small inroads into your culture. It was a continuous improvement effort. No lightning bolts. No announcements from the clouds. Just a bit-by-bit, point-by-point, experience-by-experience, concept-by-concept, tool-by-tool assault on your personal culture.

Why is this change in your personal culture so important? It's terribly basic. Remember—no one knows what business will look like in the future. This fact lies at the crux of why changing your personal culture is so important. Undeniably, like it or not, our world is changing. That change has and will continue to affect you and your safety and health program. Total quality is a process that meets these changes. Your personal culture, however, is critical to your future. New changes will occur. New ways of thinking will happen. New ways of doing things are inevitable. Without making a change in your personal culture, you might be easily swept out into this sea of change, unknowing, still protecting your own culture. By changing your culture, perhaps just a little, your vision opens to look for these changes and, hopefully, recognizes the shifts that are coming from the edge of your culture.

YOUR OPTIONS

So, upon finishing this book, you have several options. First, you can continue with "business as usual." I call it "culture protection." A bus is coming. It's called change. If you choose to remain stalwart in your culture, you're going to get hit by that bus. I know this as assuredly as I know what has occurred in the past. Change is coming. It's bigger and coming faster than ever before. If you choose not to change, watch out for that bus!

The second option you have is a personal continuum. Having changed or seriously considered changing your personal culture, you can continue to work toward a questioning-based restructuring of that culture. The key to this is having the skill to recognize paradigm shifts and the cultural maturity to challenge your own paradigms. I have a sign over my door. I see it every time I leave my office. The placement is very symbolic. The sign says, "Challenge *YOUR* Paradigms!"

It is natural for each of us to be protective of our personal culture. It defines *everything* for us. Being that important to us, it would be suicide to *not* protect it. My office represents that "safe harbor" for my culture. It is a place where my culture is unthreatened and comfortable. As I leave my office, I open myself to new experiences and information. My culture, therefore, outside my office is receptive to input.

The third option you have upon finishing this book is to accept a leadership position for change. Being a convert with a changed personal culture, you can work toward changing the culture of your organization, department, professional society or association, etc. It's an "outreach" concept. Think of it. Safety and health leading the way into the future instead of our historical way of passively watching from the sidelines. You have the option to play rather than just watch the game. You have the option to reward yourself through involvement or pay the "admission price." Through our leadership, we can revolutionize worker safety and health in America and be leaders in our organization's changes. We can ensure our futures by bringing true value to our mission and our organizations.

YOU, THE LEADER

If you're going to be a leader in cultural change, first, remember the facts about culture. You can't change facts. You must work within them if you are to be successful. Second, remember kaizen. It is not only a basic concept of changing culture, it also reflects the mind set you must have when trying to change cultures. Third, remember the importance of communication. Change cannot occur without active, multipathwayed, constant communication. It's not natural and it isn't easy, but communication must be done well. Fourth, remember the importance of participation. You are not the Lone Ranger! Only through increasing the number of forces, through increased participation, can cultural changes occur. And fifth, remember vision, mission, and planning. Don't "go boldly where no man has gone before" without having a shared vision of where you are going, knowing why you exist and what benefit your efforts provide, and having a detailed plan for getting there. As Lewis Carroll wrote in *Alice's Adventures in Wonderland*:

> Alice: "Which way do I go from here?"
> Cheshire Cat: "That depends on where you want to go."
> Alice: "I don't know where I'm going."
> Cheshire Cat: "If you don't know where you are going, any road will do."

Total quality is not a fad. It is not a buzzword that is passing through business. It is not a manufacturing concept only. It is the way to the future for business *and* for safety and health practitioners. It requires that we challenge many of our paradigms and change them. It requires that we think differently about the purity of our profession and about sharing knowledge and responsibilities. It requires dedication and commitment. From a change aspect, it is not easy. However, from a management aspect, once total quality is in place, it is extremely easy. Although total quality has many concepts and tools that run counter to our normal way of thinking, it is not complex. It works!

There are hundreds of examples of successful total quality businesses. Hundreds of books and articles have cited them. It would gain me nothing to duplicate that effort here. I cannot cite many examples of success-ful total quality safety and health programs. We've talked about why this is. It doesn't have to be that way, though. If total quality can provide the same success to your safety and health program that it has for so many businesses, why would anyone not try it? If total quality can provide so many advantages for safety and health practitioners regarding time, efficiency, productivity, mission, and performance evaluations by measurement, why would anyone choose not to embrace it?

Remember the story I used about the boy in the tree? The storm is coming. For a lot of us, it is already here. The limbs are beginning to move in the winds of change. We cannot stop the winds. We cannot calm the storm. The choice is yours. You can hope that your limb doesn't break and send you to the ground. You can change your safety and health program into a total quality program, shore up your branch, and hope your tree survives the storm. Or, you can lead the change to total quality and assure that your tree, your program, and your organization not only survive the winds of change, but continue to grow and be successful.

Choices! Throughout life we are faced with choices. Here's another one! It's your choice. Recognize that it may be the most important choice you can make. Ted Knight said it best in the movie "Caddie Shack" when he was impatiently awaiting a young golfer's decision to win or lose. He said, "Wellllll?" What will your choice be?

REFERENCES

D. Ambrose, *Leadership, The Journey Inward*. Dubuque, Iowa: Kendall/Hunt Publishing, 1991.

J.A. Barker, *Discovering the Future: The Business of Paradigms*. Burnsville, Minn.: Charthouse Learning Corp., 1989.

W.F. Christopher, *Vision, Mission, Total Quality: Leadership Tools for Turbulent Times*. Cambridge, Mass.: Productivity Press, 1993.

C.A. Coonradt, *The Game of Work*. Salt Lake City, Utah: Shadow Mountain Press, 1985.

P.F. Drucker, *Management: Tasks, Responsibilities, Practices*. New York: Perennial Management Library, Harper and Row Publishers, 1974.

R. Fukuda, *CEDAC: A Tool for Continuous Systematic Improvement*. Cambridge, Mass.: Productivity Press, 1986.

E.M. Goldratt and J. Cox, *The Goal, A Process of Ongoing Improvement*, revised edition. Croton-on-Hudson: N.Y.: North River Press, 1986.

M. Greif, *The Visual Factory, Building Participation Through Shared Information*. Cambridge, Mass.: Productivity Press, 1991.

H. Hirano, *5 Pillars of the Visual Workplace: The Sourcebook for 5S Implementation*. Cambridge, Mass.: Productivity Press, 1994.

M. Imai, *Kaizen: The Key to Japan's Competitive Success*. New York, NY: Random House, 1986.

J. Ishiwata, *IE for the Shop Floor: Productivity Through Process Analysis*. Cambridge, Mass.: Productivity Press, 1986.

J.M. Kouzes, and B.Z. Posner, *The Leadership Challenge*. San Francisco, CA: Jossey-Bass Publishers, 1987.

E. Kubler-Ross, *On Death and Dying: What the Dying Have to Teach Doctors, Nurses, Clergy and their Own Families*. New York: MacMillan Publishing, 1969.

T.S. Kuhn, "The Structure of Scientific Revolutions." *International Encyclopedia of Unified Science*, Vol. 2, No. 2. Chicago: University of Chicago Press, 1962.

D.J. Lu, trans., *Kanban, Just-In-Time at Toyota*. Japan Management Association, ed. Cambridge, Mass.: Productivity Press, 1985.

B.H. Maskell, *Performance Measurement for World Class Manufacturing*. Cambridge, Mass.: Productivity Press, 1986.

B.H. Maskell, *New Performance Measures*. Cambridge, Mass.: Productivity Press, 1993.

R. Maurer, *Feedback Toolkit: 16 Tools for Better Communication in the Workplace*. Cambridge, Mass.: Productivity Press, 1993.

S. Mizuno, *Management for Quality Improvement*. Cambridge, Mass.: Productivity Press, 1988.

Ed. Nakajima, S., *TPM: Implementing Total Productive Maintenance*. Cambridge, Mass.: Productivity Press, 1982.

Nikkan Kogyo Simbun, Ltd. and Factory Magazine, Ed. *Poke-Yoke, Improving Product Quality by Preventing Defects*. Cambridge, Mass.: Productivity Press, 1985.

T. Ohno, *Workplace Management*. Cambridge, Mass: Productivity Press, 1988.

K. Ozeki, and T. Asaka, *Handbook of Quality Tools*, Cambridge, Mass.: Productivity Press, 1990.

A. Robinson, Ed. *Continuous Improvement: A Systematic Approach to Waste Reduction*. Cambridge, Mass.: Productivity Press, 1987.

Sanno Management Development Research Center, *Vision Management: Translating Strategy into Action*. Cambridge, Mass.: Productivity Press, 1986.

E.E. Scheuing, *The Power of Strategic Partnering*. Cambridge, Mass.: Productivity Press, 1992.

S. Shiba, A. Graham, and D. Walden, *A New American TQM: Four Practical Revolutions in Management*. Cambridge, Mass.: Productivity Press, 1993.

K. Suzaki, *The New Manufacturing Challenge, Techniques for Continuous Improvement*. New York: The Free Press, Macmillian, New York, 1987.

G.H. Watson, *The Benchmarking Workbook, Adapting Best Practices for Performance Improvement*. Cambridge, Mass.: Productivity Press, 1986.

Westinghouse Quality College, Productivity and Quality Center. *Conditions of Excellence for Quality*. Pittsburgh, Penn.: Westinghouse Electric Corp., 1985.

J.N. Williamson, ed. *The Leader-Manager*. New York: Wilson Learning, John Wiley & Sons, 1986.

GLOSSARY

Benchmarking—A process of modeling aspects or parts of other programs for improving your own.

CEDAC®—Cause and Effect Diagram with the Addition of Cards: a problem-solving process where possible solutions to a problem are placed on cards, attached to a fishbone diagram and each analyzed until appropriate solutions to the problem can be identified. A registered trademark of Productivity Inc.

Challenging—The aspect of leadership that questions common or historical practice and thinking.

Change Triggers—Identifiable levels of need that may bring about a movement toward change.

Continuous Improvement—A process in which small changes are instituted that collectively and in a steady fashion improve the program or organization.

Control Program—A formal or electronic-type program that assures that each repetition of an action is accomplished the same way without oversight or intervention.

Culture—That non-formal, unwritten aspect of any individual, group, or organization that defines how it relates internally and externally. Often called the "unwritten rules."

Cycle Time—The amount of time (clock or calendar) that it takes to complete some defined action, procedure, practice or process.

Empowering—The aspect of leadership that brings the full abilities out of people.

Empowerment—The environment where each individual can achieve at the maximum level possible for the betterment of the individual, program and organization.

Enabling—The aspect of leadership that helps, provides resources, listens, and motivates others to achieve.

Encouraging—The aspect of leadership that rewards effort and celebrates accomplishment.

Japanese Management—Often called participative management, a form of managing people in which teams are encouraged to mentally and physically contribute to the mission of a program or organization.

Just-in-Time (JIT)—A process in which products or services are provided or produced in a manner to be just-in-time to meet the customers expectations and needs.

Kanban—A Japanese word that means a symbol that indicates need or readiness.

Kaizen—A Japanese word that means continuous improvement.

Leadership—The ability to manage others in such a way as to bring out the best in each member of the group and to accomplish the maximum as a group.

Line Stop—A process where a program, manufacturing process or operation is totally shut down when an error or problem is identified so that the all energies can be directed to solving the problem.

Mission Statement—A written statement that defines for a specific team, group, program or organization why it exists and provides value.

Modeling—An aspect of leadership that displays in practice the values of the leader. Often called "walking the talk."

Ownership—A situation in which an individual has known benefit in an organization's effort or goal.

Paradigm—A mental pattern, belief, or expectation that defines what is acceptable, known, or will be successful to an individual, group, or organization.

Paradigm Shift—A radical change in paradigm.

Poke-Yoke—A Japanese term that means to fix or correct something in such a manner that it can never recur.

Proactive Management—A method of management where daily activities are the result of a planned course of action toward a goal.

Process Mapping—An analytical tool in which a process is broken down into its individual steps and functions from initiation to completion.

Program Evaluation—A systematic, routine process where a program is compared to some standard of excellence.

Q-Time—That portion of cycle time that is spent waiting.

Quality Control—Traditionally, an oversight function where a product or service is inspected against a quality standard or specification.

Reactive Management—A method of management where daily activities are the result of reacting to needs outside the specific function.

SMED—A Japanese term that stands for Single Minute Exchange of Dies.

Strategic Planning—A long-term planning process that maneuvers resources and people toward accomplishing an extended goal or vision.

Theory X Management—A style of management that is highly dictatorial and autocratic.

Theory Y Management—A style of management that nurtures and responds to group decisions and actions.

Theory Z Management—Also called Japanese Management, a style of management that encourages team dynamics and decision making.

Total Quality—A method of management that is highly participative and systematically seeks improvement toward excellence.

Total Quality Management (TQM)—Managing in a total quality manner.

Transition Management—A highly demanding period of change management where a change has been started but is not complete.

Vision—A clear knowledge of where a program needs to go or what it should become to be successful in the future.

Wastes—Anything that is inefficient or wastes resources or time.

Work Cell—The arrangement of a work area that is most efficient for accomplishing its intended purpose.

X-Matrix—A process where nonreporting programs or parts of an organization can organize efforts for accomplishing a common goal or effort.

RECOMMENDED READING

Achieve International Zenger-Miller Associates Staff, et. al., *Firing on All Cylinders: The Service-Quality System for High-Powered Corporate Performance*, Business One Irwin, 1992.

Akao, Y., ed., *Hoshin Kanri: Policy Deployment for Successful TQM*, Productivity Press, 1991.

Bader, B.S., *Rediscovering Quality*, Bader Associates Inc., 1992.

Ballman, G., et al., eds., *International Service Quality & Total Quality Management Resource Guide & Directory, 1993-1994, A Comprehensive Guide to Global Resources for Facilitating Continuous Improvement*, 1993, Lakewood Publications.

Barkley, B.T., *Customer-Driven Project Management: A New Paradigm in Managing Total Quality*, McGraw, 1993.

Barnett, A. and Barnett, J., *International Interactions in Quality Management*, American Intercultural, 1993.

Batten, J., *Building a Total Quality Culture*, Gerould, W.P., Ed., Crisp Publications, 1992.

Bechtell, M.L., *Untangling Organizational Gridlock: Strategies for Building a Customer Focus*, ASQC Quality Productions, 1993.

Bonacci. E.C., *Managing for Quality & Survival: A Personal Journey Toward Excellence,* Distr. by Book World Services, Linvale Publishers, 1992.

Bowles, J. and Hammond, J., *Beyond Quality*, Berkley Publications, 1992.

Brocka, B. and Brocka, M.S., *Quality Management: Implementing the Best Ideas of the Masters*, Business One Irwin, 1992.

Brunetti, W.H., *Achieving Total Quality: Integrating Business Strategy and Customer Needs*, Quality Resources, 1993.

Caroselli, M., *Total Quality Transformations: A Resource Guide for Implementing Total Quality Training*, Human Resources Development Pr., 1991.

Caroselli, M., *Total Quality Transformations: Optimizing Missions, Methods, and Management,* Human Resources Development Pr., 1991.

Chantico Press Staff, *Total Quality Management*, QED Information Sci., 1991.

Conference Board Inc., Staff & Hiam, A., *Closing the Quality Gap" Lessons from America's Leading Companies*, P-H, 1992.

Conti, T., *Building Total Quality: A Guide for Management*, Chapman & Hall, 1993.

Conway Quality Inc. Staff, *Waste Chasers: A Pocket Companion to Quality and Productivity*, Conway Qual., 1992.

Cound, D.M., *A Leaders Journey to Quality*, Dekker, 1991.

Crosby, P.B., *Completeness: Quality for the Twenty-First Century*, NAL-Dutton, 1992.

Crouch, J.M., *An Ounce of Application is Worth a Ton of Abstraction: A Practical Guide to Implementing Total Quality Management*, Business One Irwin, 1993.

Demming, W.E., *Out of Crisis*, MIT, 1986.

Dennison, T., ed., *Zen Leadership: The Human Side of Total Quality Team Management*, Mohican Pub., 1992.

Dreger, J.B., *Primer on Total Quality Management*, Van Nos Reinhold, 1992.

Drummond, H., *The Quality Movement: What Total Quality Management is Really All About*, Nichols Pub., 1992.

Ebei, K.E., *Achieving Excellence in Business: A Practical Guide to the Total Quality Transformation Process*, Dekker, 1991.

Field, D., *Quality Malpractices*, ASQC Quality Pr., 1992.

Flood, R.L., *Beyond TQM*, Wiley, 1993.

Foulkes, S. and Stubanvoll, S., *Accelerated Systems Development*, P-H, 1992.

Gitlow, H.S., *Managing for Quality: Integrating Quality and Business Strategy*, Business One Irwin, 1992.

Goal-QPC Staff, *The Source: A Total Quality Management Information Guide*, Goal-QPC, 1991.

Greene, R.T., *Encyclopedia of Quality: How to Implement Proven Programs That Produce Measurable Results*, Business One Irwin, 1993.

Hagan, J., *Management of Quality: Strategies to Improve Quality and The Bottom Line*, Business One Irwin, 1993.

Hand, M. and Plowman, B., eds., *Quality Management Handbook*, Butterworth-Heinemann, 1992.

HMSO Staff, *Quality Management Library*, HMSO UK, 1992.

Hoffnerr, G. and Nader, G., *Breakthrough Thinking in Total Quality Management*, P-H, 1994.

Howe, R.J., et al., *Quality on Trial: Is Your Quality Initiative Paying Off?*, Quality Inst, Intl., 1992.

Hradesky, J.L., *Total Quality Management Handbook*, McGraw, 1994.

nt, V.D., *Quality in America: How to Implement a Competitive Quality Program*, Business One Irwin, 1991.

tchins, D., *Achieve Total Quality*, P-H, 1991.

land, L.R., *Quality Management in Projects and Programs*, Proj. Mgnt. Inst., 1991.

ikawa, K., *What is Total Quality Control?*, P-H, 1991.

lonski, J.R., *Implementing Total Quality Management: An Overview*, Pfeiffer & Co., 1991.

olementing TQM: Competing in the Nineties Through Total Quality Management, Technical Management Consortium, 1992.

inson, H.T., *Relevance Regained: From Top-Down Control to Bottom-Up Empowerment*, Free Press, 1992.

inson, R.S., *Management Processes for Quality Operations*, Quality Pr., 1993.

inson, R.S. and Kazense, L.A., *The Mechanics of Quality Processes*, ASQC Qual. Pr., 1993.

ilaw, D.C., *Continuous Improvement and Measurement for Total Quality*, Business OneIrwin, 1992.

rkham, R.L., *A Better Way: Achieving Total Quality*, American Training Alliance, 1992.

yoshi, U., et al., *TQM for Technical Groups: Total Quality Principles for Product Development*, Productivity Press, 1993.

ox, D., *Effective Organizational Change: The Strategic Planning Process*, Ican Pr., 1993.

issoff, L.L., *Closing The Gap: The Handbook for Total Quality Implementation*, LLK Associates, 1992.

bovitz, G., et al., *Making Quality Work: A Leadership Guide for the Results-Driven Manager*, Harper Business, 1993.

m, K.D., *Total Quality: A Textbook of Strategic Quality Leadership and Planning*, Air Acad. Pr., 1992.

wler, E.E., *Employee Involvement and Total Quality Management: Practices and Results in Fortune 1000 Companies*, Jossey-Bass, 1992.

ary, B. and MacDorman, J., *Quality Manager's Handbook*, AT&T Customer Info., 1992.

Loden, M and Rosener, J.B., *Workforce America! Managing Employee Diversity As A Vital Resource*, Business One Irwin, 1990.

Logothetis, N., *Managing for Total Quality*, P-H, 1992.

MacDermott, R., et al., *Employee-Driven Quality: Releasing the Creative Spirit of Your Organization Through Suggestion Systems*, Quality Resc., 1993.

MacDonald, J., *Understanding TQM: In a Week*, NTC Pub. Grp., 1993.

MacDonald, J and Piggot, J., *Global Quality: The Management Culture*, Pfeiffer & Co., 1993.

McNair, C.J., *Benchmarking: Adding Distinctive Value to Every Aspect of Your Business*, Harper Business, 1992.

Madu, C.N., ed., *Management of New Technologies for Global Competitiveness*, Greenwood, 1993.

Mahoney, F.X., and Thor, C.G., *The TQM Trilogy: Using ISO9000, the Demming Prize and the Baldrige Award to Establish a System for Total Quality Management*, AMACOM, 1994.

Menon, H.G., *TQM in New Product Manufacturing*, McGraw, 1992.

Miller, G.L., *Whats, Whys and Hows of Quality of Improvement*, ASQC Qual. Pr., 1992.

Miller, W.C., *Quantum Quality: Quality Improvement Through Innovation, Learning and Creativity*, Qual. Resc., 1993.

Oakland, J.S., *Total Quality Management*, Butterworth-Heinemann, 1993.

Oliver Wright Publications, Inc. Staff, *The Oliver Wright ABCD Checklist for Operational Excellence*, Oliver Wright, 1992.

Osada, T., *The Five S's: Five Keys to a Total Quality Environment*, Qual. Resc., 1991.

Persico, J, Jr., ed., *The TQM Transformation: A Model for Organizational Change*, Qual. Resc., 1992.

Peters, B.H. and Peters, J.L., eds., *Total Quality Management*, Conference Board, 1991.

Picogna, J.L., *Total Quality Leadership: A Training Approach*, International Info. Assoc., 1993.

Pierce, R.J., ed., *Leadership, Perspective and Restructuring for Total Quality*, ASQC Qual. Pr., 1991.

Systems, Inc. Staff, *Total Quality Transformation Improvement Tools*, PQ
Systems, 1991.

ıdings on Total Quality Management, Assn. Natl. Advertisers., 1992.

lly, N.B., *Quality: What Are They Talking About?*, Van Nos Reinhold,
1993.

ources for the Implementation of Total Quality Management, OMS, 1992.

berts, H. and Sergesketter, B.F., *Quality is Personal: A Foundation for
Total Quality Management*, Free Press, 1993.

ss, J., *Quality Management: Text, Cases and Readings*, St. Lucie Press,
1993.

an, J.M., *The Quality Team Concept in Total Quality Control*, ASQC Qual,
pr., 1992.

hkin, M. and Kiser, K.J., *Putting Total Quality Management to Work:
What TQM Means, How to Use It and How to Sustain It Over the
Long Run*, Berrett-Kohler, 1993.

shkin, M and Kiser, K.J., *Total Quality Management*, Ducochon Press,
1991.

ılor, J., *TQM Field Manual*, McGraw, 1992.

ıaaf, D. and Kaeter, M., *Pursuing Total Quality: One Hundred One
Logical Ways to Improve Quality for Your Customers*, Lakewood
Publications, 1992.

ımidt, W.H. and Finnigan, J.P., *The Race Without a Finish Line: America's
Quest for Total Quality*, Jossey-Bass, 1992.

ıuler, R.S. and Harris, D., *Managing Quality*, Addison-Wesley, 1992.

ecter, E.S., *Managing World Class Quality: A Primer for Executives and
Managers*, Dekker, 1991.

ıder, N.H. and Dowk, J.J. Jr., *Vision, Values and Courage: Leadership for
Quality Management*, Macmillan, 1994.

ın, S.S., *Total Quality Control Essentials: Key Elements, Methodologies and
Managing for Success*, McGraw, 1992.

ıthworth, M. and Southworth, D., *Management or Continuous Quality
Improvement*, Graph Arts Pub., 1992.

endolini, M.J., *The Benchmarking Book*, AMACOM, 1992.

lley, D.J., *Total Quality Management*, ASQC Qual, Pr., 1991.

Thomas, B., *Total Quality Training: The Quality Culture and Quality Trainer*, McGraw, 1992.

Townsend, P.L. and Gebhardt, J.E., *Quality in Action: Ninety-Three Lessons in Leadership*, Wiley, 1992.

Tunks, R., *Fast Track to Quality: A Twelve-Month Program for Small to Mid-Sized Businesses*, McGraw, 1992.

Weaver, C.N., *TQM: A Step-by-Step Guide to Implementation*, ASQC Qual. Pr., 1991.

Werner, J., *Managing the Process, the People and Yourself*, ASQC Qual. Pr., 1992.

Wickman, R.F. and Doyle, R.S., *Breakthrough Quality Improvement for Leaders Who Want Results*, ASQC Qual. Pr., 1993.

Wiggenton, K., *Quality Improvement Team Helper* , AT&T Customer Info., 1990.

Wilkerson, D., et al., *Measuring Quality: Linking Customer Satisfaction to Process Improvement*, Coopers Total Quality, 1993.

Willie, R., *Leading the Quality Initiative*, AT&T Customer Info., 1990.

Winchell, W.O. intro by, *TQM: Getting Started and Achieving Results with Total Quality Management*, SME, 1994.

Worth, *Quality Management and How to Achieve It*, Inst. Econ. Finan., 1993.

INDEX

About Government Institutes

Government Institutes, Inc. was founded in 1973 to provide continuing education and practical information for your professional development. Specializing in environmental, health and safety concerns, we recognize that you face unique challenges presented by the ever-increasing number of new laws and regulations and the rapid evolution of new technologies, methods and markets.

Our information and continuing education efforts include a Video-tape Distribution Service, over 140 courses held nation-wide throughout the year, and over 150 publications, making us the world's largest publisher in these areas.

Other related books published by Government Institutes:

Safety Made Easy: A Checklist Approach to OSHA Compliance— Written by Tex Davis, this book provides a new, simpler way of understanding your requirements under the complex maze of OSHA's workplace safety and health regulations. The easy-to-use format and logical organization make this book ideal for those who are just entering the field of safety compliance as well as for experienced safety professionals. *Softcover/200 pages/May '95/$45 ISBN: 0-86587-463-8*

OSHA Field Operations Manual, 6th Edition — This step-by-step manual developed by OSHA for use by its own Compliance Safety and Health Officers in carrying out their inspections will show you where the inspectors will look, what they will look for, and how they'll evaluate your working conditions. *Softcover/456 pages/Feb '94/$85 ISBN: 0-86587-380-1*

OSHA Field Inspection Reference Manual—Be prepared in advance for your next OSHA inspection! This new revision of inspection guidelines, previously contained in the OSHA Field Operations Manual, is now being used by OSHA inspectors when checking your facility for compliance. Learn where the inspectors will look, what they'll look for, how they'll evaluate your working conditions, and how they'll actually proceed once inside your facility. *Softcover/ 144 pages/Jan '95/$59 ISBN: 0-86587-426-3*

Health & Safety Risk Management: Guide for Designing an Effective Program This practical guide provides a boilerplate system for any company, large or small, to develop a working health and safety program. A disk may also be added which contains the full text of the book to enable companies to create their own custom programs by just inserting site specific information. *Three-ring binder only/310 pages/ June '94/$225 ISBN: 0-86587-397-6 Three-ring binder with Microsoft Word files on disk (#4052)/or ASCII disk (#4053)/$250*

OSHA Compliance Handbook — This practical non-legalese guide, written by the law firm of Reed Smith Shaw & McClay, will put you on track toward meeting your OSHA requirements. Covers: OSHA Standards; General Duty Clause; Recordkeeping; Hazard Communication; Inspections; Civil Penalties and Violations; and much more. *Softcover/ 400 pages/May '92/$83 ISBN: 0-86587-290-2*

Chemical Information Manual on Disk — Used by OSHA's inspectors as a reference for sampling chemicals during industrial hygiene inspections, this database contains essential data on over 1400 chemical substances. The disk runs under Windows,and allows you to search through the database in four different ways: Chemical name; synonyms; CAS #; and IMIS #. Revised as of January, '95, this manual is a must-have reference tool! *June '95/$99 #4070*